Virtual Humans
Today and Tomorrow

Chapman & Hall/CRC Artificial Intelligence and Robotics Series

Series Editor: Roman Yampolskiy

The Virtual Mind
Designing the Logic to Approximate Human Thinking
Niklas Hageback

Intelligent Autonomy of UAVs
Advanced Missions and Future Use
Yasmina Bestaoui Sebbane

Artificial Intelligence
With an Introduction to Machine Learning, Second Edition
Richard E. Neapolitan, Xia Jiang

Artificial Intelligence and the Two Singularities
Calum Chace

Behavior Trees in Robotics and AI
An Introduction
Michele Collendanchise, Petter Ögren

Artificial Intelligence Safety and Security
Roman V. Yampolskiy

Artificial Intelligence for Autonomous Networks
Mazin Gilbert

Virtual Humans
Today and Tomorrow
David Burden, Maggi Savin-Baden

For more information about this series please visit:
https://www.crcpress.com/Chapman--HallCRC-Artificial-Intelligence-and-Robotics-Series/book-series/ARTILRO

Virtual Humans

Today and Tomorrow

David Burden

Maggi Savin-Baden

CRC Press
Taylor & Francis Group
Boca Raton London New York

CRC Press is an imprint of the
Taylor & Francis Group, an **informa** business
A CHAPMAN & HALL BOOK

CRC Press
Taylor & Francis Group
6000 Broken Sound Parkway NW, Suite 300
Boca Raton, FL 33487-2742

Printed on acid-free paper

International Standard Book Number-13: 978-1-138-55801-4 (Hardback)

Visit the Taylor & Francis Web site at
http://www.taylorandfrancis.com

and the CRC Press Web site at
http://www.crcpress.com

Suppose that sometimes he found it impossible to tell the difference between the real men and those which had only the shape of men, and had learned by experience that there were only two ways of telling them apart... first, that these automata never answered in word or sign, except by chance, to questions put to them; and second, that though their movements were often more regular and certain than those of the wisest men, yet in many things which they would have to do to imitate us, they failed more disastrously than the greatest fools.

René Descartes
The Philosophical Writings of Descartes: Volume 3, The Correspondence
(Trans. J. Cottingham., R. Soothoff., D. Murdoch., A. Kenny. Cambridge, UK: Cambridge University Press, 1991)

To our partners, Deborah and John,
the least virtual humans we know.

Permissions

The quote from *Snow Crash* by Neal Stephenson in Chapter 2 is used with permission of Penguin Random House and Darhansoff & Verrill Literary Agents.

Contents

List of Figures

List of Tables

Acknowledgements

THIS BOOK HAS BEEN a challenge to create in these liquid times of moving technologies. We are grateful to colleagues and family for keeping us up to date and informed, as the initial reviewers of the proposal, and to Randi Cohen at CRC Press/Taylor & Francis Group, who has been so positive about it as a project. Any mistakes and errors are ours; we are after all, only physical humans.

Authors

David Burden has been involved in artificial intelligence (AI), virtual reality (VR) and immersive environments since the 1990s. David set up Daden Limited in 2004 to help organisations explore and exploit the social and commercial potential of using chatbots, AI and virtual environments. David and his team have delivered over 50 immersive learning and chatbot projects for clients across the globe and have led over a dozen collaborative research projects funded by InnovateUK and the Ministry of Defence (MOD). David was a finalist in British Computer Society's (BCS) machine intelligence competition and has authored over 20 papers on virtual worlds and AI. In his early life, David was an officer in the Royal Corps of Signals, and he still maintains an interest in wargaming, alongside science-fiction role-playing (*Traveller*) and space exploration – although he's just as happy walking and trekking on high hills or in remote places.

Maggi Savin-Baden is Professor of Education at University of Worcester and has researched and evaluated staff and student experience of learning for over 20 years and has gained funding in this area (Leverhulme Trust; JISC; Higher Education Academy; MoD). She has a strong article record of over 50 research articles and 16 books, which reflects her research interests on the impact of innovative learning, digital fluency, cyber-influence, pedagogical agents, qualitative research methods, and problem-based learning. In her spare time, she runs, bakes, climbs, and attempts triathlons.

Introduction

Virtual Humans are typically seen as human-like characters on a computer screen or speaker with embodied life-like behaviours, which may include speech, emotions, locomotion, gestures and movements of the head, eyes or other parts of an avatar body. At one end of the scale are smart-speaker and virtual assistant systems, such as Siri, Alexa and Cortana, the chatbot-based virtual coaches found in several mobile phone applications and the customer service chatbots which are becoming increasingly prevalent on the Internet. More developed examples include virtual tutors, virtual life coaches, digital twins and virtual personas, and even the non-player characters in some computer games. At the other, more sophisticated, end of the scale are the virtual humans of science fiction, from *Iron Man*'s J.A.R.V.I.S and *Red Dwarf*'s Holly to *Star Trek Voyager*'s Emergency Medical Hologram. All portray themselves as human, and offer differing degrees of human capability, and, indeed, humanness.

This book argues that virtual humans are a phenomenon of the twenty-first century, and while some people have debated their impact on society, few have researched them in-depth. It presents an overview of current developments, research and practice is this area thus far, offering a fair and honest overview of the issues, outlining the risks, not exaggerating the claims but providing evidence and research. It will review current practices at a time when education is changing rapidly with digital, technological advances. It will outline the major challenges faced by today's developers, teachers and researchers, such as the possibility for using virtual humans for teaching, training and practice. The book will situate many of the discussions around related applications with which the general reader may be aware (for example, Siri, Cortana and Alexa, as already mentioned), and draw examples from speculative and science fiction (from *Pygmalion* and *Prometheus*, in classical mythology, to *Battlestar Galactica*, *Her* and Channel 4's *Humans*).

The book is not particularly concerned with the use by physical humans of avatars in, for example, virtual worlds (another use of the term virtual human). However, such situations can put the virtual human and physical human on an equal footing within a virtual environment.

The first section of the book, *Part I: The Landscape,* outlines understandings of virtual humans and begins by providing much needed definitions and a taxonomy of artificial intelligence. This section includes *Chapter 1: What Are Virtual Humans?*, which presents an introductory analysis of the traits which are important when considering virtual humans, argues for a spectrum of virtual human types, and evaluates the common virtual human forms against that spectrum. The second chapter, *Chapter 2: Virtual Humans and Artificial Intelligence*, broadens this discussion by engaging with the issue that there is a range of perspectives about what counts as artificial intelligence (AI) and virtual humans. Thus, this chapter presents a virtual humans/AI landscape and positions the different terms within it, as well as identifying three main challenges within the landscape that need to be overcome to move from today's chatbots and AI to the sort of AI envisaged in the popular imagination and science fiction literature.

The second section of the book, *Part II: Technology*, explores approaches to developing and using relevant technologies, and ways of creating virtual humans. It presents a comprehensive overview of this rapidly developing field. This section begins with *Chapter 3: Body*, which examines the technologies which enable the creation of a virtual body for the virtual human, explores the extent to which current approaches and techniques allow a human body to be modelled realistically as a digital avatar and analyses how the capability might develop over the coming years. The chapter also explores the senses, and examines how human (and other) senses could be created for a virtual human. *Chapter 4: Mind* complements the study of the 'body' and 'senses' in Chapter 3 by examining the different technologies and approaches involved in the creation of the mind or 'brain' of a virtual human. The chapter will consider current research in the areas of perception, emotion, attention and appraisal, decision-making, personality, memory, learning, and meta-cognition, and anticipates the future directions that may develop over the coming decade.

The next chapter, *Chapter 5: Communications*, explores how the senses and abilities created for a virtual human can support language and non-verbal communications. *Chapter 6: Architecture* then reviews some of the leading architectures for creating a virtual human. The range considered will show the breadth of approaches, from those which are theoretical, to those that have emerged from being engineering-based, to those inspired by neuroscience. The final chapter in the section, *Chapter 7: Embodiment*, begins by examining the current and significant research around the concept of embodiment, and some of the challenges to it. The importance of 'grounding' will also be discussed. The possibilities for the use of virtual worlds to provide that space for embodiment and grounding will then be presented, and its implications for virtual human development assessed. *Chapter 8: Assembling and Assemblages* considers how all the elements considered so far can be used to build a virtual human and describes some of the more common types of virtual human which could be encountered now or in the future. It then uses several different lenses to assess the 'humanness' of such virtual humans, and whether a virtual human needs to be more than the sum of its parts.

Part III: Issues and Futures is the final section of the book, which examines issues such as identity and ways of dealing with the complexities and ethical challenges of these liquid technologies. It also examines possible futures, including digital immortality. It begins with *Chapter 9: Digital Ethics* which explores some of the ethical, moral and social issues and dilemmas which virtual humans introduce. It also examines what it means to undertake ethical research in the context of virtual humans and examines issues, such as technical ethics, design ethics, legal issues and social concerns, including honesty, plausibility and the nature of consent. The next chapter, *Chapter 10: Identity and Agency*, explores the notion of virtual humans from a broad perspective by examining studies into identity and agency in virtual reality, immersive environments and virtual worlds.

Chapter 11: Virtual Humans for Education examines the current and possible future impact that virtual humans could have on education and analyses the potential impact of virtual humans on education in terms of new developments and uses of virtual humans for guiding and supporting learning. Building on the issues of relationships, *Chapter 12: Digital Immortality* explores the emotional, social, financial and business impact that active digital immortality could have on relations, friends, colleagues and institutions. This chapter presents recent developments in the area of digital immortality, explores how such digital immortals might be created and raises challenging issues but also reflects on the ethical concerns presented in Chapter 9. The final chapter, *Chapter 13: Futures and Possibilities*, considers what the impact of virtual humans might be, and what significant related developments might take place, within three successive timeframes: 2018–2030; 2030–2050; and 2050–2100 and beyond. The chapter then examines how the three main challenges to the developments of virtual humans identified in Chapter 2, namely improving humanness, Artificial General Intelligence (AGI) and Artificial Sentience (AS) might be addressed in order to move from virtual humanoids to 'true' virtual humans, and even virtual sapiens.

This book focusses on the virtual human technologies that are available currently or will be available by 2020. It is not concerned with more speculative concepts, such as brain interfaces and mind uploads, 'evil AIs' and the technological singularity, although some of these will be discussed in Chapter 13. This book is also not concerned with robots, i.e., physical bodies controlled by computer programs. Instead, it is suggested here that any suitably developed virtual human should be able to slip in and out of a robot body in the same way as it might take control of an avatar, and the unique challenges of robotics are fundamentally electro-mechanical ones.

With existing technologies, it is possible to create some form of proto-virtual human, a virtual humanoid which can exhibit some of the characteristics of a physical human, and may even deceive people in some areas, but which, as a holistic digital copy of a physical human, falls short.

However, it is only through starting to build and create virtual humans that are more than simple chatbots or personal assistants that researchers, engineers, ethicists, policymakers and even philosophers will be able to investigate and begin to understand the true challenges and opportunities that the creation of fully realistic and perhaps sentient virtual humans might create. If such virtual humans do become a possibility during this current century, then their impact on human society will be immense.

virtualhumans.ai Website

Supporting material for this book, including links to virtual human images, videos, applications and related work and papers, can be found on the website at www.virtualhumans.ai.

I

The Landscape

INTRODUCTION

Part I sets the bounds to this book. Many similar terms are used for artefacts, which are, to some greater or lesser extent, virtual or digital versions of a physical human. These range from chatbots, conversational agents and autonomous agents to virtual humans and artificial intelligences. There is also a blurring between the digital and physical versions of virtual humans, the latter being represented by robots and androids. Somewhere between the two sit the digital entities which are linked to specific physical platforms such as Siri and Alexa.

Chapter 1 will consider some of these different manifestations of virtual humans and identify a set of traits which can be used separate virtual humans from other software systems, and to compare and contrast different versions of virtual humans. This leads to a working definition of a virtual human and also to the identification of lower-function virtual humans – termed virtual humanoids, and higher function virtual humans, which are the true equivalent of physical humans and have been termed virtual sapiens. The chapter closes with an examination of several key use cases of virtual humans which will be considered throughout the book, namely those of chatbots, conversational agents and pedagogic agents.

Chapter 2 will consider the broader software landscape within which virtual humans exist. In particular, the chapter will look at the landscape of artificial intelligence, which, in the late 2010s, is a term which is being much abused and being applied to everything from data analytics to self-driving cars. This landscape will be used to help place the concepts of virtual humanoids, virtual humans and virtual sapiens, and to identify the three big challenges facing virtual human development. Chapter 2 will conclude by looking at how virtual humans are represented in the popular media – on film, TV, in books, on the radio, in both computer and roleplaying games and even on stage. Whilst science-fiction cannot give much guidance on how to actually build a virtual human, it can help identify what it would be like to live and work with them, highlight moral and ethical issues, and even give some useful insights into possible ontologies and terminology.

CHAPTER 1

What Are Virtual Humans?

INTRODUCTION

Virtual Humans are human-like characters, which may be seen on a computer screen, heard through a speaker, or accessed in some other way. They exhibit human-like behaviours, such as speech, gesture and movement, and might also show other human characteristics, such as emotions, empathy, reasoning, planning, motivation and the development and use of memory. However, a precise definition of what represents a virtual human or even 'artificial intelligence' (AI) is challenging. Likewise, establishing the distinctions between different types of virtual human, such as a chatbot, conversational agent, autonomous agent or pedagogic agent is unclear, as is how virtual humans relate to robots and androids. This chapter presents an introductory analysis of component parts of a virtual human and examines the traits that are important when considering virtual humans. It examines existing definitions of a virtual human before developing a practical working definition, and argues for a spectrum of virtual human types, and presents some common examples of virtual humans.

WHAT IS A VIRTUAL HUMAN

A virtual human is, fundamentally, a computer program. In the far future, it may be something else, but for the foreseeable future, a virtual human is simply code and data which has been designed, and may be evolving, to give the illusion of being human.

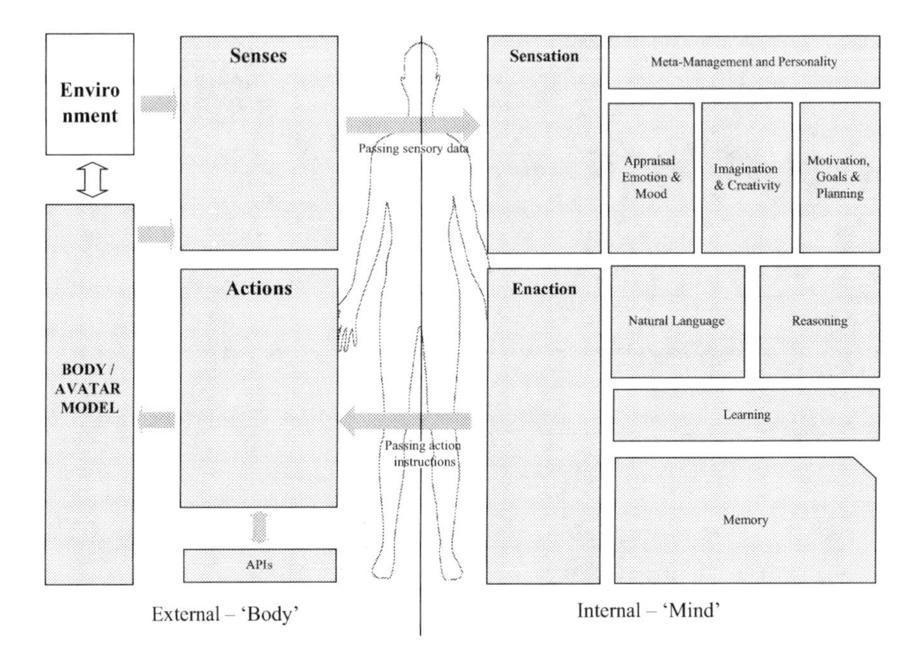

FIGURE 1.1 Elements of a virtual human.

Figure 1.1 provides a very basic schema of the possible elements of a virtual human program and its environment. Not all of these elements would be expected to be present in every virtual human (as will be seen, some are very challenging to implement), and not every element needs to be present to the same degree. The diagram will be developed more fully in Part II and all the component parts considered in some detail.

The main elements are:

- A body, which may be a digital avatar or simply a microphone and speaker or text-chat interface
- A set of senses and the ability to detect sensations
- The ability to appraise sensation triggers and respond to them, including showing emotion and changing mood
- The ability to plan to achieve goals, ideally set by some internal motivation
- The ability to reason and problem solve
- The ability to show imagination and creativity

- The ability to communicate in natural human language
- The ability to learn
- The ability to remember and access memories
- The ability to manage all the above, which may be reflected in a personality
- The ability to enact its decisions through taking actions with its 'body'
- Application Programming Interfaces (APIs) to other systems
- An environment in which to exist and interact with

EXISTING DEFINITIONS OF VIRTUAL HUMAN

There are many definitions of 'virtual human' in the literature, and just a few are considered here. Rickel defines virtual humans as autonomous agents that support face-to face interaction in virtual environments (Rickel et al., 2002). The difficulty with this definition is that it is unclear what an autonomous agent or a virtual environment is, and implies a relatively limited scope of action, and an emphasis on only face-to-face interaction. The definition is more appropriate to a non-player-character within a computer game than recognising the wider and more capable forms that a virtual human could take.

Chatbots.org, a leading hobbyist website on chatbots, identifies 161 different terms for human-like conversational agents (Van Lun, 2011). The website defines virtual humans as automated agents that converse, understand, reason and exhibit emotions. They possess a three-dimensional body and perform tasks through natural language-style dialogues with humans.

This offers a broader range of capabilities for the virtual human, and the explicitness of face-to-face interaction has been replaced by a more general 'natural language' definition and the need to be in a virtual environment has been removed.

Traum defines virtual humans as artificial agents that include both a visual body with a human-like appearance and range of observable behaviours, and a cognitive component that can make decisions and control the behaviours to engage in human-like activities (Traum, 2009).

This definition moves beyond the focus on dialogue and interaction, and highlights that a broader range of human activities and behaviours may be involved.

FROM SELF-DRIVING CAR TO VIRTUAL HUMAN

A useful way to help consider what is and is not a virtual human is to think of a modern self-driving car. Such a car will have sophisticated sensors (senses), complex decision-making algorithms acting on a sub-second basis and dealing with, ostensibly, moral issues, for example, to stay on course and collide or to swerve and risk hitting a mother and baby. It will also have a high level of autonomy within its domain, for example, the 'driver' chooses where to go, but the car chooses the route and is controlling the steering wheel. However, while the car is unlikely to be seen as a virtual human, it may be called intelligent (indeed, this field of research is often known as 'intelligent vehicles'). The question, though, is what happens if the car is given a natural language interface for defining the route and discussing road conditions with speech input and output? It then appears to be more human, but possibly a long way from David Hasselhoff's KITT car in the *Knight Rider* TV series. With the addition of some virtual assistant capability, such as access to your diary and emails, and access to the web and Wikipedia, it begins to be more than just a car, and a lot closer to the KITT model. From there, it would be possible to create a personality, emotion and motivation, along with a head-and-shoulders animated avatar on the pop-up screen in the car so that you have a *Red Dwarf* Holly-like character to whom you can relate and talk. Yet what then changes if an android robot is placed in the driver's seat? It's a purely mechanical device slaved to the car's virtual human-like interface—but there is now not only the issue about whether the car and the character are separate, but also whether the humanness resides in the car or the android. Perhaps, it needs the android to get out of the car and say 'Drive yourself' for us to consider it truly human!

As this example shows, the boundaries between what is and what is not a virtual human are fluid, as is whether it is a good (accurate, effective, useful) or bad (inaccurate, ineffective, useless) virtual human. However, the traits described below should help to identify just how 'virtual human' something is.

THE TRAITS OF A VIRTUAL HUMAN

In defining a virtual human, it is useful to examine the different traits which could be used to differentiate a virtual human from other forms of computer program that may be either showing intelligence or are based on some human characteristic. In some cases, such traits may relate directly to the elements shown in Figure 1.1, but others may be more holistic. The traits considered here, and often presented as dichotomies, are:

- Is it physical or digital?
- Is it manifest in a visual, auditory or textual form?
- Is it embodied or disembodied?
- Is it humanoid or non-humanoid?
- Does it use natural language or command-driven communication?
- Is it autonomous or controlled?
- Is it emotional or unemotional?
- Does it have a personality?
- Can it reason?
- Can it learn?
- Is it imaginative?
- How self-aware is it?

Physical or Digital?

The first of these traits is whether the entity is defined by a physical or digital presence. In much popular literature, for example, replicants in *Blade Runner* or Kryten in *Red Dwarf*, physical androids are considered virtual humans. Whilst possessing a presence in some form of humanoid or non-humanoid physical robot body may well be useful at times to a virtual human, the essence of the virtual human is in its digital form. This is particularly salient when, represented as an avatar within a virtual world, it is able to present itself as just as 'human' as any avatar controlled by a physical human.

It should be noted that many authors refer to software-based virtual humans as 'digital humans' (for example, Jones et al., 2015 and Perry, 2014), but the term 'digital humans' is also used in other areas, such as filmmaking (Turchet et al., 2016) and ergonomic design (Keyvani et al., 2013) to refer only to the creation of the 'body' of the virtual human, and not of any higher functions. Hence, there is a preference in this work for the term 'virtual human.'

Visual, Auditory or Textual?

Having defined the virtual human as essentially a digital construct, it is questionable as to whether it matters how the virtual human is manifest within the digital domain. For example, the virtual human could be presented as a two-dimensional (2D) head-and-shoulders animated avatar (for example, the Sitepal system – www.sitepal.com), as a three-dimensional (3D) avatar within a virtual world, such as 'Second Life' (www.secondlife.com), as a voice on the phone, home speaker or a Skype call, or as a participant within a text-based chat room. It is argued here that the virtual human should be independent of its digital (or physical) manifestation. Indeed, it is the possibility for fluidity between different forms that could be an important capability for the future. However, in order to be recognised as a virtual human, the entity must have some ability to communicate through one of these, or closely related modes. Yet, even the most basic computer program can have a text interface, so having such interfaces is not a sufficient condition to define a virtual human (Figure 1.2).

Embodied or Disembodied?

Whilst closely linked to the issue of manifestation, there is also the matter of whether the virtual human needs to be embodied, digitally or physically. One belief in cognitive science is that 'intelligence' needs to be embodied (Iida et al., 2004); it needs the sensation, agency and grounding of having a 'body' within a rich and changing environment in order to develop and experience 'intelligence'. Whilst these issues will be discussed in more detail in Chapter 7, there does appear to be a case that an entity which never has some form of embodied manifestation may never be able to become a 'true' virtual human—although it could still be a very smart computer or artificial intellect, a so-called artilect. If being embodied is a requirement of a virtual human, the implication would be that many of the 'virtual human' computers of science fiction (Hal, Orac, J.A.R.V.I.S, Holly) would be better considered as artilects, not virtual humans.

(a) (b)

FIGURE 1.2 (a) 2D and (b) 3D virtual human avatar.

Humanoid or Non-Humanoid?

The next concern is whether the entity has a humanoid form. Myth and popular literature contains numerous examples of humans able to take on animal and other forms, such as lycanthropes and Native American skin-walkers; so just because a virtual human can represent itself as something other than human does not mean that it cannot be a virtual human. If the entity's core personality, default appearance, actions and thought processes are those of a human, then it should be considered as a virtual human. However, if the entity only ever represented itself in a particular animal form in appearance, deed and thought, then it should be termed a virtual animal.

Natural Language or Command-Driven Communication?

A key capability for almost any human is the ability to communicate through 'natural language', be that through voice, signing or other mechanisms. This implies a relatively free-flowing conversation using accepted vocabulary and grammar, and with the ability to track the context of the conversation and make appropriate responses to the utterances of others. Indeed, it is often seen as a sign of mental impairment, whether from mental health problems, such as dementia or a temporary state, like drunkenness, when a person is not able to communicate to this standard. The free-flowing conversation requirement precludes the inclusion systems based on simple command-based exchanges. So, to be considered a virtual human, an entity must be able to do more than merely respond to a set of commands. There are gray areas where commands give way to natural language. Entities such as Siri and Alexa could readily be thought of as virtual humans despite their natural language conversational ability being limited, since the natural, human, inclination towards anthropomorphism can readily see them as 'human'. The Turing Test (Turing, 1950), and related Loebner Prize (Floridi et al., 2009), which will be discussed in more detail in Chapter 2, have become useful ways of evaluating the natural language conversational ability of chatbots. Yet, it is unlikely that having a good natural language capability, as in a good chatbot, is a sufficient condition for being a virtual human, although it may well be a necessary one.

Autonomous or Controlled?

Despite some of the current philosophical and neurological debates about freewill (for example, Caruso, 2013), it is generally accepted that a human has autonomy for most practical purposes, although such autonomy is limited by laws and social, moral, and ethical frameworks. A true virtual human should also therefore be expected to have a similar degree of autonomy and be bound by similar frameworks. Whilst a virtual human may not initially exhibit the same level of autonomy as a physical human, its level of autonomy should still be extensive within the scope of its coding.

A further consideration is how much intrinsic motivation the virtual human possesses. Autonomy often exists within a well-defined set of tasks, for instance, a self-driving car choosing a route and then a more complex set of second-by-second decisions over speed and direction, based on a rapidly evolving and complex environment, need to be made. Once the journey is completed, the car just waits for its next command or possibly drives itself off home to its garage to recharge. A true virtual human, though, should always be operating in a relatively autonomous way. Once it finishes one task, it needs to decide on its next one. Such decisions should be driven by a set of long-term goals, by motivation, as they are in physical humans.

Emotional or Unemotional?

Demonstrating and responding to emotions is certainly seen in popular culture as being evidence of humanity (for example, the Voight-Kampff test in *Blade Runner*), and the lack of emotions is often taken as an indication of a disturbed or even psychotic personality. Certainly, within the literature (for example, Mell, 2015; Mykoniatis, 2014), the ability for a virtual human to be able to show and respond to emotions is seen as an important feature. One of the key questions is to what extent the virtual human is 'faking it' – does it 'feel' emotion or empathy – or is it just exhibiting the features and responses that we associate with those traits? Often, though, it is the emotional response of the human party to the virtual human's condition that can be just as important (for example, Bouchard et al., 2013). So, if an emotional reaction is elicited in a physical human to a virtual human showing emotion, then do the mechanisms through which that emotion was generated matter? Since surely almost any emotion portrayed on stage or in film is artificial as well?

Presence of a Personality?

In considering the possible traits of a virtual human, the word 'personality' is often used. Personality can be defined in a variety of ways, and personality theories include: dispositional (trait) perspective, psychodynamic, humanistic, biological, behaviourist, evolutionary and those based on social learning. This suggests that a virtual human should also appear to behave, feel and think in a unique and individual way, not identical to any other virtual human (except, possibly, clones of itself) or even to any other physical human. If the virtual human does not show a unique personality, or indeed any personality, then it is possibly not worthy of the term. However, there is again the danger of anthropomorphism, people will quite readily attribute personalities to very obviously non-human objects, from cars to printers, and it is important to be cautious about whether any perceived personality is just being implied by the observer or is actually present within the system.

Ability to Reason?

Reasoning here is used to refer to the ability to accept a set of inputs and make a sensible decision based on them. Reasoning can include theories about moral reasoning, as suggested by Kohlberg (1984), as well as models of novice to expert reasoning used in professional education (Benner, 1984). Reasoning is also taken here as including problem-solving, which is a more constrained version of the reasoning ability. At its lowest level within a virtual human, the reasoning may be as simple as identifying that if a website customer has enquired about lighting, then they should be shown all the desks, tables and floor lamps in stock. In a more developed virtual human, it would be expected that the reasoning capability is beginning to match that of a human – in other words, given the same inputs, it will make a similar decision to a human, even though the number of factors, or their relation to the output, might be more complex, or there may be high degrees of uncertainty involved, so called fuzzy or even wicked problems.

Can It Learn?

One common definition of an intelligent system is that of having the capacity to learn how to deal with new situations (Sternberg et al., 1981). Intelligence is not so much about having the facts and knowledge to answer questions (a more popular view of what intelligence is), but rather an adaptive ability to cope with new situations, often by applying patterns and knowledge (reason) previously acquired. As such, the ability to learn (in a whole variety of different ways and applied to a whole variety of different situations) must be an important trait for a virtual human.

One of the ultimate goals of AI research is so-called Artificial General Intelligence (discussed in more detail in Chapter 13), a computer system that exhibits a very generalized and human form of such learning and adaptability. For current practical purposes, it could be expected that a virtual human would show some ability to learn and adapt within the scope of its programming.

Is It Imaginative?

As will be discussed in Chapter 4, there are a lot of computer programs which demonstrate creativity, using parametrics, neural networks, genetic algorithms or other approaches to create pieces of music, paintings, poems or other works of art. There is, however, a difference between creativity and imagination. The imagination trait is more about an internal ability to visualise something, something which may not exist or at least has not been sensed, and perhaps to take an existing trope and change its parameters to create a whole new experience. The 'creative' element is then more about taking this piece of imagination and using craft, skills and 'creativity' to make it manifest and bring it into the social domain. So, the important trait is probably that of imagination, with creativity coming from combining imagination with other traits, such as reasoning (what colour where) and learning (how did I do this last time).

Sentient or Non-Sentient?

In common discourse, sentience can be viewed as an equivalence of 'thinking': Does the machine have a cognitive function? Is it sentient? There is also some potential overlap with free will and autonomy. A further definition of sentience aligns it with consciousness, but defining that is similarly fraught with problems. Indeed, the question of what consciousness means in terms of the way that we have subjective, phenomenal experiences is often described as the 'hard problem'. Such consciousness implies some form of self-awareness and internal narrative, for example Nagel's 'What Is It Like to Be a Bat?' (Nagel, 1974). Achieving sentience using current technologies is beyond the present capabilities for a virtual human. However, an aspiration to develop some form of internal self-awareness and internal narrative and dialogue would seem to be desirable, and whether that results in, or can enable, some form of true sentience is probably a key philosophical and research question for our times.

DEMONSTRATING INTELLIGENCE?

It should be noted that in the 10 traits above, 'intelligence' has been deliberately omitted. There is no real agreement as to what an 'intelligent' system is, just as there is no agreement about what counts as human intelligence.

Intelligence has been defined as everything from logic to leadership, and from emotional knowledge to problem solving. Indeed Legg and Hutter (2007) list 71 definitions for 'intelligence' and 'artificial intelligence'. What are taken as indicators of intelligence are considered to be well enough covered by a combination of the other traits, particularly learning and reasoning.

A VIRTUAL HUMAN PROFILE

In Figure 1.3, the 10 traits that have been described above are represented graphically. A high rating for a trait would be marked by a point towards the outer edge of the decagon, and a low rating towards the centre. Joining each point by a line (as shown in Figures 1.4 and 1.5) then provides a radar-plot or spider-diagram style profile for a particular instance or type of virtual human. It should be noted, though, that some traits need to be far more developed than others to be treated as some form of virtual human. For instance, an entity with even a relatively low level of personality, natural-language or emotion is more likely to be taken as a virtual human than one with a very high degree of reasoning or learning ability but which does not show any of the other traits. Almost any level of self-awareness would also put an entity well on the way to being thought of as a virtual human.

It should also be noted that the 'digital or physical' trait described earlier has not been added to the chart as: (a) it is a very binary trait – the

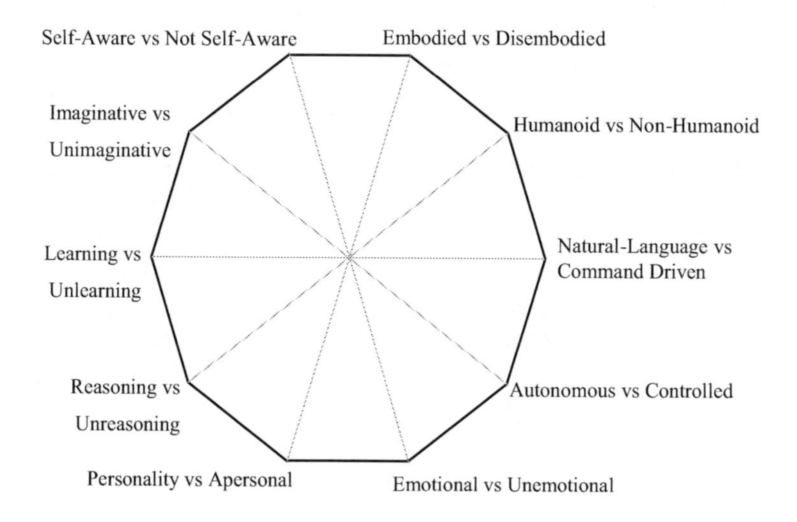

FIGURE 1.3 Traits of virtual humans.

entity is either presenting as pixels on a screen or as a lump of metal or organic material, (b) in defining virtual humans, their digital nature has already been asserted, and (c) any developed virtual human would be able to move in and out of (or rather commandeer) a physical representation, as required.

Just from the examples discussed so far, it is clear that the term 'virtual human' could represent a wide spectrum of capabilities. In its most typical manifestation, a virtual human:

- Manifests itself in a visual, auditory, textual or similar form,
- May have some embodiment within a virtual world,
- Presents itself as primarily humanoid in manifestation and behaviour,
- Will have a natural language capability,
- May exhibit a degree of autonomy,
- May have an ability to express, recognise and respond to emotions,
- May exhibit some aspects of a personality,
- May have some ability to reason in a human-like way,
- May, possibly, exhibit some elements of imagination, and
- May even have a self-narrative, but is unlikely to have any indications of sentience.

VIRTUAL HUMANOIDS AND VIRTUAL SAPIENS

To aid later analysis, it is perhaps useful to define terms for virtual humans which are biased to the simpler or more complex ends of the spectrum, so that there is a clearer understanding of what is being discussed. Two new terms are proposed.

At the lower, less functional, end of the spectrum is the 'virtual humanoid'. This could be defined as a digital entity which:

- Can manifest itself in a visual, auditory, textual or similar form,
- Need not have any sense of embodiment within a virtual or physical world,
- Is primarily humanoid in manifestation and behaviour,

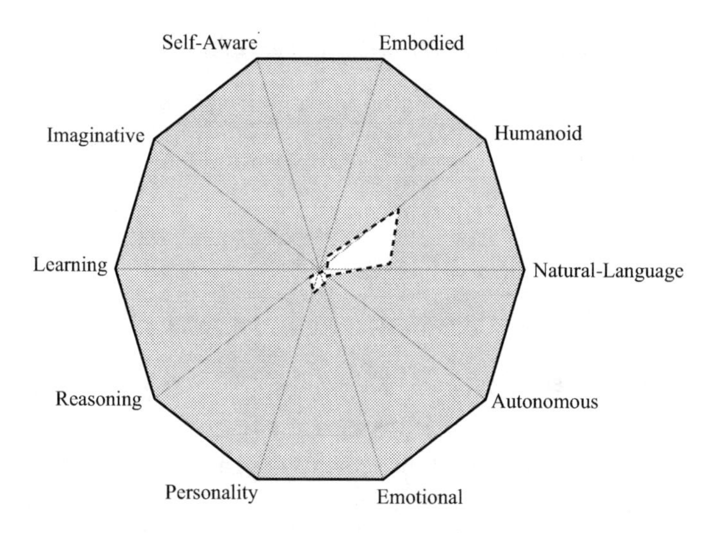

FIGURE 1.4 Profiles of virtual humanoids (dashed line), virtual sapiens (solid line – effectively the complete edge of the decagon) and virtual humans (the shaded space in between).

- Can respond to commands, but does not have to exhibit a developed natural language capability,
- Does not need to exhibit any autonomy,
- Does not need to express or recognise emotion,
- Does not need to exhibit a (unique) personality – most of its personality being implied by anthropomorphism,
- Does not need to show any significant ability to reason,
- Does not need to show any imagination, and
- Does not need to have any indications of internal narrative or sentience.

At the upper (most developed) end is the 'virtual sapien', a digital entity which:

- Can manifest itself in a visual, auditory, textual or similar form,
- Will have embodiment within a virtual or physical world,
- Is primarily humanoid in manifestation and behaviour,

- Has a highly developed natural language capability,
- Exhibits a high degree of autonomy and intrinsic motivation,
- Can express, recognize and respond to emotions,
- Exhibits a unique personality,
- Can reason in a human-like way,
- Exhibits elements of the use of imagination, and
- Has self-awareness and may have some indications of sentience.

The two new definitions are shown as profiles in Figure 1.4, with virtual human as the more overarching term.

It should be noted that the lines between a virtual humanoid, virtual humans (the all-embracing term), and virtual sapiens are significantly blurred, and on many measures, it is a matter of degree rather than of absolutes.

TOWARDS A WORKING DEFINITION

Whilst such graphics as Figures 1.3 and 1.4 can be valuable tools, it is also useful to have a simple working definition of what a virtual human is. The consideration of personality above – thinking, feeling, and behaving in an individual way – certainly goes some way to providing a more concise working definition:

> *A virtual human is a digital entity (or perhaps, more generally, a program, algorithm or even a process) which (looks) thinks, feels and behaves like a human.*

On this basis, a virtual humanoid could be considered as a digital entity which just looks and, to a certain extent, behaves like a human, whilst virtual sapiens is a digital entity which could pass for a human in an extended unrestricted evaluation.

There is indeed a very strong temptation to choose a behaviouristic definition of a virtual human, typified, perhaps, by Turing's original 'imitation game' (Turing, 1950). Behaviourist definitions have, though, been frequently challenged (for example, Searle, 2014), and Chalmer's arguments about philosophical zombies (Chalmers, 1996) explore similar issues.

So, perhaps, more useful working definitions are:

- *Virtual Humanoids* – Simple virtual humans which present, to a limited degree, as human and which may reflect, in a limited way, some of the behaviour, emotion, thinking, autonomy and interaction of a physical human.
- *Virtual Humans* – Software programs which present as human, and which may have behaviour, emotion, thinking, autonomy and interaction modelled on physical human capabilities.
- *Virtual Sapiens* – Sophisticated virtual humans which achieve similar levels of presentation, behaviour, emotion, thinking, autonomy, interaction, self-awareness and internal narrative to a physical human.

Note: The term digital entity has been used above, but perhaps a stricter and more general definition would be an informational entity, as it then avoids the limitation on form that 'digital' (and also program) could imply—for example, ruling out some biological possibilities and even intelligent windows, as in *Permutation City* (Egan, 1994). The term 'infomorph' (Muzyka, 2013) has been used for such an informational entity. The term 'artilect', introduced earlier, would then represent a relatively well developed infomorph.

EXAMPLES OF VIRTUAL HUMANS

Having set the scope for what represents a virtual human, it is useful to look at some examples of virtual humans placed more at the virtual humanoid end of the spectrum, which can be encountered in today's world. It is also helpful to consider how the terms used for them, such as chatbots, autonomous agents, conversational agents and pedagogical agents, should be understood against the definitions and profiles developed above. Further examples will be considered in Chapter 8.

Chatbots

Chatbots is a generic term for describing a piece of software that mimics human conversation. It emphasises the conversational capability but says nothing about any other elements of the virtual human. A system like Siri or Alexa that does not really engage in conversation is probably not even a chatbot, rather being a question-answering or

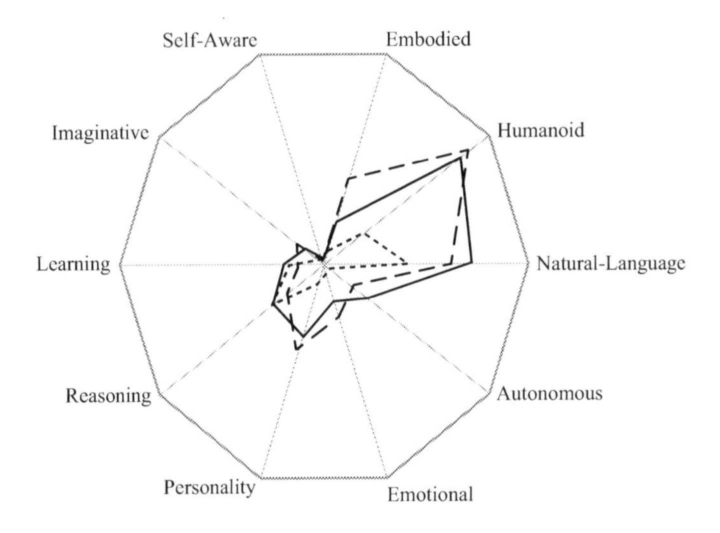

FIGURE 1.5 Mapping current virtual human types: chatbots (dashed), conversational agents (dotted), and pedagogical agents (solid).

command-taking system, but would fit within the virtual humanoid definition. Importantly, the term chatbot does not imply anything about sentience or intelligence, it is just a system attempting to mimic human conversation.

Chatbots can be thought of as rudimentary pieces of software that aim to create the illusion of a human. They rely on a pre-existing corpus of information that is used to respond to human questions and stimuli. Basic chatbots do not have the ability to identify the progressions of a conversation, nor to adapt responses based on previous answers given by the chatbot or as input by the human user, and nor may they aim to assist in the completion in any particular task. A simple chatbot aims to identify each human input (through pattern matching or keyword searching) and respond with the appropriate answer retrieved from a store of responses with little processing of context. Examples of this kind of chatbot can be found incorporated into toys, such as 'Hello Barbie' (Rhodan, 2015), which allows children to have pseudo conversations with a doll.

Autonomous Agents

Autonomous agent is a very broad term that has been used for well-developed chatbots, with elements of the virtual human (Bogdanovych, 2005), for massively replicated software entities running in a crowd simulation (Pelechano et al., 2007), and for software programs carrying out very specific but autonomous tasks—for example, stock trading (Subramanian et al., 2006). As such, the use of the term in relation to virtual humans is perhaps best avoided.

Conversational Agents

Conversational agents (Cassell, 2000) are virtual humans whose task is to provide a conversational interface to a particular application, rather than through command line instructions or the clicking of icons or menu items, for tasks ranging from making travel bookings and buying furniture to interrogating sales and marketing data.

These 'agents' can be represented textually, orally or in conjunction with an avatar (or all three), and, like chatbots, may also exhibit some of the behaviours and characteristics of humans, such as speech, locomotion, gestures and movements of the head, eye or other parts of the body (Dehn and Van Mulken, 2000). However, their role goes beyond simply maintaining a conversation with no particular goal. The level of sophistication of these types of agent can thus determine their utility within differing contexts.

Across the services, manufacturing and raw materials sectors of industry, conversational agents have been used to increase the usability of devices and as methods to assist the retrieval of information. For example, numerous websites have utilised agents as virtual online assistants to improve access to information, such as 'Anna' on the Ikea website or the shopping assistants from H&M or Sephora accessed through the KiK service (KiK Interactive Inc., 2016).

Similarly, smartphone and tablet interfaces now include options to use personal assistant applications such as 'Siri', 'Cortana' or 'Google Now' to assist users in searching for information, starting applications and performing routine tasks, such as sending messages. Other software, such as Amazon's 'Alexa', incorporated into their 'Echo' device, allow the control of various household smart devices such as fridges and heating systems. These agents, while being helpful, do not necessarily have a high level of contextual awareness, and so, whilst being effective at assisting with simple tasks, such as information retrieval or acting on simple commands (for example, Alexa), agents are not always positioned to provide guidance.

Conversational agents are likely to possess a defined goal and set of capabilities. Granting them additional open-ended conversational capabilities, emotions or motivations could be counterproductive for such agents in achieving their goal, and, therefore, damaging to their owner. Most conversational agents are towards the lower end of the virtual humans' spectrum, although they may have well-developed natural language capabilities; they are invariably, but not exclusively, more task orientated than chatbots.

Pedagogical Agents

Virtual humans used in education are more commonly referred to as pedagogical agents, or sometimes more narrowly, as virtual tutors. The function of pedagogical agents is to aid learners during the learning process. These agents aim to support learners by providing easier access to relevant information and to improve motivation (for example, Bowman, 2012; Schroeder and Adesope, 2014). As will be discussed later in this book, the sophistication of an agent dictates its role and the types of interactions achievable with humans.

The literature presents a number of examples of pedagogical agents that have been used successfully in learning and educational contexts. However, these have been predominantly focused on agents that may not have an understanding of dialogue or of the progression of conversations between the individual and the agent. Hasler, Touchman and Friedman (2013) found that in a comparison of human interviewees with virtual world pedagogical agents (in non-learning situations), pedagogical agents and human interviewees were equally successful in collecting information about their participants' real-life backgrounds. Pedagogical agents, being neither human interviewees nor text-based surveys, therefore pose an interesting opportunity for the educator seeking to facilitate student discussion.

Virtual Mentors

Virtual Humans can have varying levels of perceived understanding of any topic, which can be adequate for low level tasks, such as helping to access databases of knowledge. However, a deeper understanding of the content of any discussion between user and virtual human, combined with increased personalisation options, can allow the virtual human to act as a mentor and provide guidance, rather than merely simple access to information. There are many examples of such virtual mentors, particularly in the mobile space (where the interaction can seem particularly intimate and personal) including Wysa (https://www.wysa.io/) and Woebot (https://woebot.io/).

Figure 1.5 shows how some of these examples map onto the virtual human profile.

ARTIFICIAL INTELLIGENCE, MACHINE LEARNING AND VIRTUAL HUMANS

There does appear to be a current tendency to attribute any 'clever' computer algorithm to be an example of artificial intelligence. It was IBM's success with Watson in the Jeopardy challenge in 2011 that appears to have started this trend. In popular literature, an AI is usually taken to mean a high-functioning virtual human or other software entity, with signs of sentience – referred to here as 'virtual sapiens'. Media and press reports often suggest that AI is more developed than it actually is, and it is important to be able to distinguish between clever algorithms, some form of well-developed virtual human, virtual sapiens and what might be, ultimately, termed as an artificial sentience This will be considered in more detail in Chapter 2.

Discussions around artificial intelligence often bring in the concept of machine learning and, in particular, neural networks. Machine learning is defined here as a computer system that can learn to make decisions based on the examination of past inputs and results, so that its future decisions optimise some parameter, such as recognising faces in photographs. Whilst a machine learning system could well be *part* of a virtual human, it is certainly insufficient to be a complete one. If the term 'machine learning' is actually used to mean a neural network-based system, then it is probably not a necessary one either.

CONCLUSION

This chapter has sought to establish the bounds of what is meant by a virtual human and identify some key traits which can be used to identify the differing forms and capabilities of virtual humans. A key point is that a virtual human operates primarily within the digital domain, although it may sometimes have access to a robot body. The term virtual humanoid has been introduced for a low-functioning virtual human (more akin to simple chatbots and conversational agents), and the term virtual sapiens for the highest level of virtual humans that show signs of self-awareness and 'sentience'. Examples of virtual humans in the forms of chatbots, conversational agents, pedagogic agents and virtual mentors have been described.

In Chapter 2, the relationship between virtual humans and 'artificial intelligence' will be explored in more detail, and the ways in which virtual humans have been portrayed in the media and arts will be surveyed to help better understand the public perception of virtual humans.

REFERENCES

Benner, P. (1984). *From Novice to Expert: Excellence and Power in Clinical Practice*. Boston, MA: Addison-Wesley Publishing.

Bogdanovych, A., Simoff, S., Sierra, C., & Berger, H. (2005). Implicit training of virtual shopping assistants in 3D electronic institutions. In *Proceedings of the IADIS International e-Commerce 2005 Conference*, December 15–17, 50–57, Porto, Portugal: IADIS Press.

Bouchard, S., Bernier, F., Boivin, E., Dumoulin, S., Laforest, M., Guitard, T., Robillard, G., Monthuy-Blanc, J., & Renaud, P. (2013). Empathy toward virtual humans depicting a known or unknown person expressing pain. *Cyberpsychology, Behavior, and Social Networking, 16*(1), 61–71.

Bowman, C. D. D. (2012). Student use of animated pedagogical agents in a middle school science inquiry program. *British Journal of Educational Technology, 43*(3), 359–375.

Caruso, G. (Ed.). (2013). *Exploring the Illusion of Free Will and Moral Response.* New York: Lexington.

Cassell, J. (2000). *Embodied Conversational Agents.* Cambridge, MA: MIT press.

Chalmers, D. J. (1996). *The Conscious Mind: In Search of a Fundamental Theory.* Oxford, UK: Oxford University Press.

Dehn, D. M., & Van Mulken, S. (2000). The impact of animated interface agents: A review of empirical research. *International Journal of Human-Computer Studies, 52*(1), 1–22.

Egan, G. (1994). *Permutation City.* London, UK: Orion.

Floridi, L., Taddeo, M., & Turilli, M. (2009). Turing's imitation game: Still an impossible challenge for all machines and some judges—an evaluation of the 2008 Loebner contest. *Minds and Machines, 19*(1), 145–150.

Iida, F., Pfeifer, R., Steels, L., & Kuniyoshi, Y. (Eds.) (2004). Embodied artificial intelligence. *Lecture Notes in Computer Science*, 3139.

Jones, A., Unger, J., Nagano, K., Busch, J., Yu, X., Peng, H. Y., Alexander, O., Bolas, M., & Debevec, P. (2015). An automultiscopic projector array for interactive digital humans. In *ACM SIGGRAPH 2015 Emerging Technologies* (SIGGRAPH '15), New York: ACM.

Hasler, B. S., Tuchman, P., & Friedman, D. (2013). Virtual research assistants: Replacing human interviewers by automated avatars in virtual worlds. *Computers in Human Behavior, 29*, 1608–1616.

Keyvani, A., Lämkull, D., Bolmsjö, G., & Örtengren, R. (2013). Using methods-time measurement to connect digital humans and motion databases. In *International Conference on Digital Human Modeling and Applications in Health, Safety, Ergonomics and Risk Management* (pp. 343–352). Berlin, Germany: Springer.

Kohlberg, L. (1984). *The Psychology of Moral Development: The Nature and Validity of Moral Stages (Essays on Moral Development, Volume 2).* New York: Harper & Row.

Legg, S., & Hutter. M. (2007). A collection of definitions of intelligence. *Frontiers in Artificial Intelligence and Applications*, 157, 17–24. Available online https://arxiv.org/pdf/0706.3639.pdf.

Mell, J. (2015). Toward social-emotional virtual humans. In *Proceedings of the 2015 International Conference on Autonomous Agents and Multiagent Systems*. Istanbul, Turkey: International Foundation for Autonomous Agents and Multiagent Systems.

Muzyka, K. (2013). The outline of personhood law regarding artificial intelligences and emulated human entities. *Journal of Artificial General Intelligence, 4*(3), 164–169.

Mykoniatis, K., Angelopoulou, A., Proctor, M. D., & Karwowski, W. (2014). Virtual humans for interpersonal and communication skills' training in crime investigations. In Shumaker R., and Lackey S. (Eds.) *Virtual, Augmented and Mixed Reality. Designing and Developing Virtual and Augmented Environments* (pp. 282–292). VAMR 2014. Lecture Notes in Computer Science, vol 8525. Cham, Switzerland: Springer.

Nagel, T. (1974). What is it like to be a bat? *The Philosophical Review, 83*(4), 435–450.

Pelechano, N., Allbeck, J. M., & Badler, N. I. (2007). Controlling individual agents in high-density crowd simulation. In *Proceedings of the 2007 ACM SIGGRAPH/Eurographics Symposium on Computer Animation* (pp. 99–108). Aire-la-Ville, Switzerland: Eurographics Association.

Perry, T. (2014). Leaving the uncanny valley behind. *IEEE Spectrum, 51*(6), 48–53.

Rickel, J., Marsella, S., Gratch, J., Hill, R., Traum, D., & Swartout, W. (2002). Toward a new generation of virtual humans for interactive experiences. *IEEE Intelligent Systems, 17*(4), 32–38.

Rhodan, M. (2015). This advocacy group is saying 'Hell No' to 'Hello Barbie'. *Time*. Available online http://time.com/4093660/barbie-hell-no/.

Schroeder, N. L., & Adesope, O.O. (2014). A systematic review of pedagogical agents' persona, motivation, and cognitive load implications for learners. *Journal of Research on Technology in Education, 46*(3), 229–251.

Searle, J. R. (2014). Introduction: Addressing the hard problem. *Journal of Integrative Neuroscience, 13*(2), 7–11.

Sternberg, R. J., Conway, B. E., Ketron, J. L., & Bernstein, M. (1981). People's conceptions of intelligence. *Journal of Personality and Social Psychology, 41*(1), 37–55.

Subramanian, H., Ramamoorthy, S., Stone, P., & Kuipers, B. J. (2006). Designing safe, profitable automated stock trading agents using evolutionary algorithms. In *GECCO '06 Proceedings of the 8th Annual Conference on Genetic and Evolutionary Computation* (pp. 1777–1784). (GECCO '06). New York: ACM.

Traum., D. (2009). Models of culture for virtual human conversation. In C. Stephanidis (Ed.). *Proceedings of the 5th International Conference on Universal Access in Human-Computer Interaction. Part III: Applications and Services* (UAHCI '09), Berlin, Germany: Springer-Verlag.

Turchet, F., Romeo, M., & Fryazinov, O. (2016). Physics-aided editing of simulation-ready muscles for visual effects. In *ACM SIGGRAPH 2016 Posters* (p. 80). ACM.

Turing, A. M. (1950). Computing machinery and intelligence. *Mind, 59*(236), 433–460.

Van Lun, E., (2011). 161 Humanlike conversational AI synonyms. Chatbots.org. Available online https://www.chatbots.org/synonyms/.

CHAPTER 2

Virtual Humans and Artificial Intelligence

INTRODUCTION

Many early researchers thought of the goal of artificial intelligence (AI) research as being the creation of a human-like, generally applicable intelligence and were focused on real-world tests to prove it, such as the Turing Test. With the AI Winter (Hendler, 2008) of the late 1970s and 1980s, much AI research switched to expert systems. In the last decade, the focus has been on 'narrow-AI' – driverless cars, machine learning techniques and so on. In both these cases, the potential commercialization of the technology has been a driving force, and the 'blue-sky' research for some far distant science-fiction level AI has taken a backseat.

One of the challenges of discussing virtual humans and AI, is that everyone appears to have a different idea about what counts as AI. It is also evident from the literature that the term virtual humans tends to be used as an overarching term that includes other terms, such as chatbots, conversational agents and pedagogical agents, as discussed in Chapter 1. To structure the discussion, this chapter will present an AI landscape and position the different types of virtual humans and AI within it. More significantly, the chapter will also identify the three main challenges within the landscape that need to be overcome to move from today's chatbots and AI to the sort of AI envisaged by those early researchers, and in the popular imagination and science fiction literature.

AN ARTIFICIAL INTELLIGENCE LANDSCAPE

In the 1970s, there used to be a saying that 'what we call artificial intelligence is basically what computers can't do yet'. Activities that were thought to require intelligence, such as playing chess, when mastered by a computer ceased to need 'real' intelligence. In the late 2010s, the situation appears to be the opposite and to read most press releases and media stories it now appears that 'artificial intelligence' is basically anything that a computer *can* do. In order to get a better definition of what 'artificial intelligence' is (at least in the context of this book) a simple model of the computing landscape has been developed and is shown at Figure 2.1. Almost any computer program can be plotted on it, with the vast majority of programs occupying the smallest part of the bottom-left-hand corner.

The two axes will now be considered in turn.

Complexity and Sophistication

The bottom axis of Figure 2.1 illustrates complexity and sophistication of the program. Figure 2.2 shows this axis in more detail. Four main points

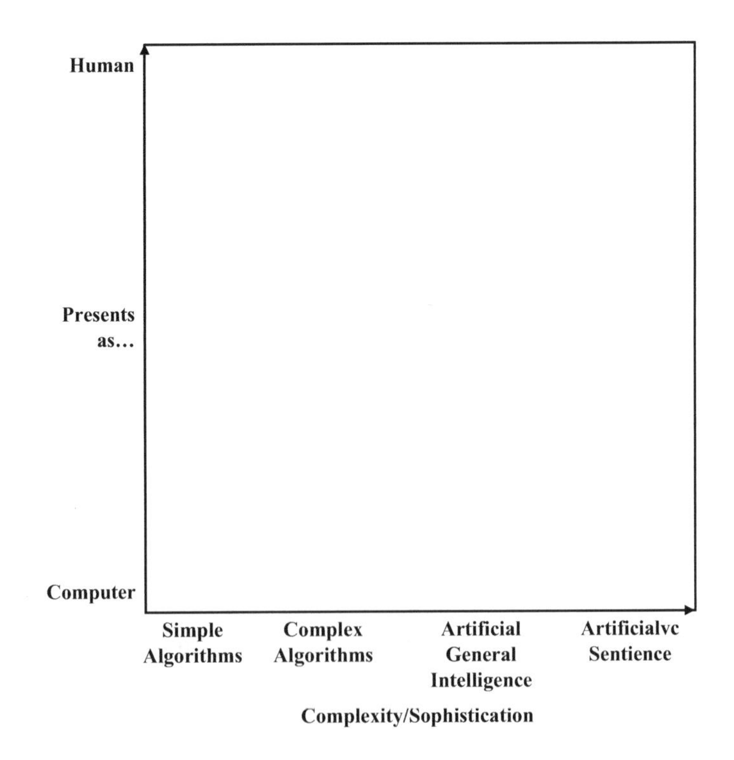

FIGURE 2.1 The AI/computing landscape.

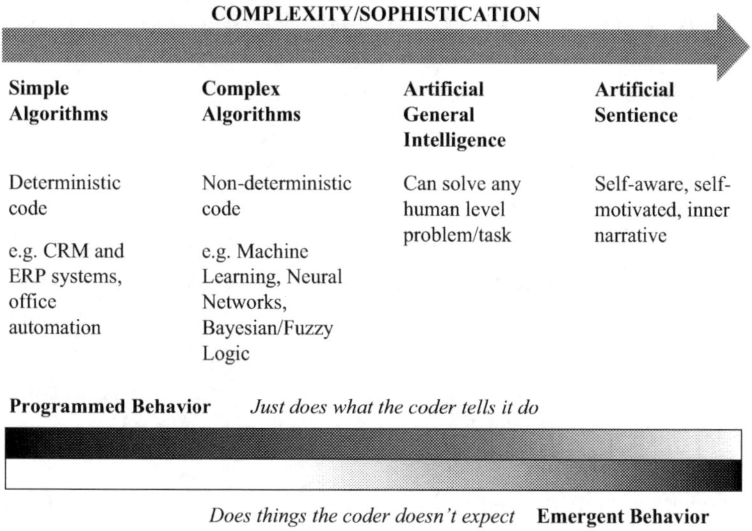

COMPLEXITY/SOPHISTICATION

Simple Algorithms	**Complex Algorithms**	**Artificial General Intelligence**	**Artificial Sentience**
Deterministic code	Non-deterministic code	Can solve any human level problem/task	Self-aware, self-motivated, inner narrative
e.g. CRM and ERP systems, office automation	e.g. Machine Learning, Neural Networks, Bayesian/Fuzzy Logic		

Programmed Behavior *Just does what the coder tells it do*

Does things the coder doesn't expect **Emergent Behavior**

FIGURE 2.2 The complexity/sophistication dimension.

on the axis are presented, although the axis is a continuum, and boundaries are fluid and overlapping:

- **Simple Algorithms** – probably 99% of most computer programs, even complex Enterprise Resource Planning (ERP) and Customer Relationship Management (CRM) systems since they are highly linear and predictable.
- **Complex Algorithms** – programs such as, but not limited to, machine learning, deep learning, neural networks, Bayesian networks and fuzzy logic where the complexity of the inner code starts to move beyond simple linear relationships. Many systems currently referred to as AI sit here.
- **Artificial General Intelligence** – closer to what the public image of AI is, a system that can be applied to a wide range of problems and solve them to a similar level as a human.
- **Artificial Sentience** – beloved of science-fiction, code which 'thinks' and is 'self-aware', the possibility and implications of which will be discussed later in this book.

Systems that reach the level of an Artificial General Intelligence (AGI) or artificial sentience (AS), as discussed later in Chapter 13, would be accepted by most people as being called AI, but the term is also applied to lesser systems. Part of the reason for that is in the vertical axis where the more human seeming a system is the more people are inclined to believe it has intelligence.

Presentation and Humanness

The second dimension, shown in the vertical axis in Figure 2.1, is about 'presentation' – does the program present itself as human, or indeed another animal or being, or as a computer – just a collection of windows, menus and forms using a graphical user interface (GUI) or even an old-fashioned command-line interface. However, even an application with a command-line interface can appear 'intelligent' if it is fronted in the form of a text-based chatbot which can answer questions on data or access and control other systems on demand.

The position of a system on the vertical axis reflects the program's ability to present in several dimensions of 'humanness', as shown in Table 2.1.

Each of these, and other contributing elements of 'humanness' will be considered in more detail in Part 2 of this book.

Marketing AI versus Real AI

In Figure 2.3, the shaded area on the chart illustrates the area occupied by current, so-called AI technologies and some examples of where some specific systems, such as chatbots, computer game Non-Player-Characters (NPCs), personal assistants, expert systems and autonomous cars, are located.

TABLE 2.1 Presentational Dimensions of Humanness

Text-to-Speech	**Does it sound human?**
Speech Recognition	Can it interpret human speech?
Natural Language Understanding	Can it understand and respond appropriately to natural language?
Natural Language Generation	Can it create its own sentences or does it use templated responses?
Avatar Body Realism	Does its body look real?
Avatar Face Realism	Does its face look real?
Avatar Body Animation	Does it move in a realistic way?
Avatar Expression	Does its face make realistic expressions and do its lips move appropriately with any speech?
Emotion	Does it show appropriate emotional responses?
Empathy	Can it detect the emotional state of other people and respond appropriately?

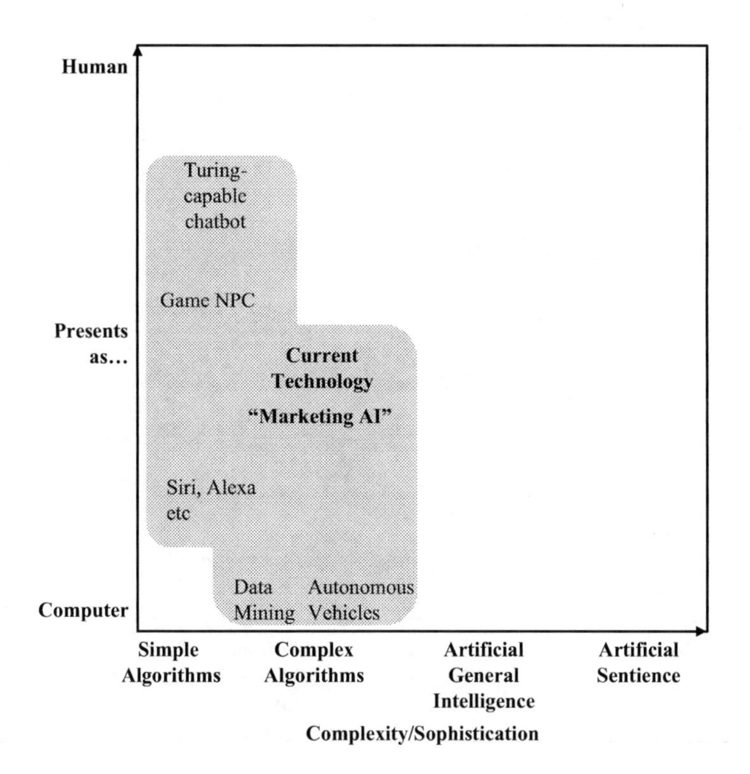

FIGURE 2.3 Marketing AI.

As can be seen, the shaded area occupied by the marketing and media use of the term AI covers only a small part of the chart. As Lea (2017) suggests:

> Because many useful algorithms are machine intelligence derivatives, it seems almost any algorithm can be labelled as AI, regardless of whether or not it actually is, and regardless of how trivial that algorithm may be. Marketing AI is used to dress up both ordinary computer programs and systems, and otherwise dull acquisitions or commercial ventures.

This can be compared with the more populist, science-fiction derived, view of what AI should be, as shown in the darker shaded area of Figure 2.4. Some specific examples will be described in more detail later in this chapter, but it is evident that these systems are occupying very different areas on the chart, with not even any overlap.

Perhaps a more useful way to describe the current uses of AI, especially in the low humanness/moderate sophistication area might be 'automated intelligence'. This describes what a lot of this sort of AI is trying to achieve, by

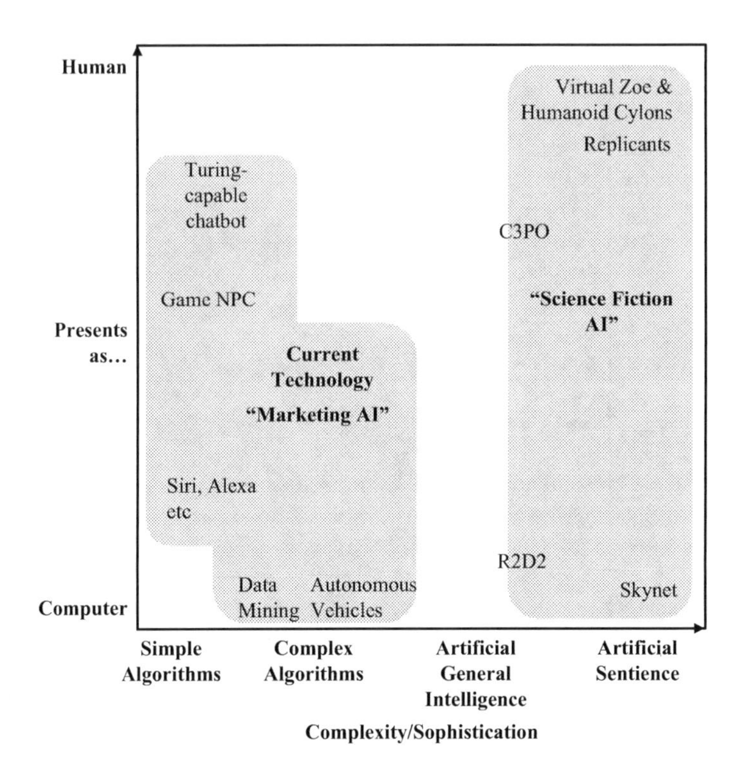

FIGURE 2.4 The populist/science fiction view of AI.

taking one aspect of human intelligence and automating it, but with minimal regard for any aspect of humanness, from self-driving cars to marketing driven data mining. The use of 'automated' also suggests that there are some limits to the abilities. Others have also referred to this more constrained form of AI as 'narrow AI' (Pennachin and Goertzel, 2007). The phrase 'machine intelligence' is also often used synonymously with AI, for example, in the British Computing Society's own Machine Intelligence competition.

THE THREE BIG CHALLENGES IN AI DEVELOPMENT

In order to grow the space currently occupied by today's AI, there are two possible directions to move in; making the entities seem more human or making them more complex and 'intelligent'. In achieving these goals, there are three challenges to be faced:

- Improving Humanness.
- Increasing Generalization.
- Realizing Sentience.

Much of this book is dedicated to understanding how far we are in overcoming the first challenge, and, in Chapter 13, to considering how the second two challenges might be addressed.

THE TURING TEST AND UNCANNY VALLEY

The Turing Test, and the concept of the uncanny valley, represent useful ways to evaluate how far up the vertical axis a system is, and to what extent it is addressing the challenge of humanness.

The Turing Test as first postulated by Alan Turing (Turing, 1950) was meant as a test for machine intelligence based on whether a user sitting at two computer terminals could tell on which one they were conversing with a human, and on which with a computer. The test has been much discussed ever since (for example, Saygin et al., 2000; Epstein et al., 2009). Whilst it is no longer seen as test of 'intelligence', it is certainly a useful, if not ideal, test of chatbots and natural language capability. It will be discussed in more detail in this context in Chapter 5. However, within the virtual human sense, it is often useful to think of a version of the test for other characteristics, from simple appearance, such as telling images of real and computer-generated faces apart, to more complex behaviours, such as deciding which avatar showing empathic behaviour is being controlled by a human and which is being controlled by a computer. A system which is able to pass such a Turing Test for a particular capability is often described as being 'Turing-capable'.

The Loebner Prize (Bradeško and Mladenić, 2012) has been an annual implementation of the Turing Test that offers a Bronze award for the best chatbot that still fails to convince any judges of its humanness, a Silver award for a chatbot which passes the Turing Test in text-chat only, and a Gold award for a chatbot which passes the Turing Test in speech and video (Powers, 1998).

Mori's concept of the 'uncanny valley' (Mori, 1970) relates to how objects which are almost, but not quite human, create more of a sense of unease than those which are decidedly not human. Mori identified that when considering artificial human forms, humans tolerated, but did not particularly emotionally respond to industrial robots, but then began to show affinity and empathy for toy robots and good quality (anime and manga style) human-like but still visually artificial toys. However, their affinity fell as they encountered things like prosthetic hands and some actors' masks which were more human in one way but caused more unease in the viewer. Many computer avatars sit squarely in this uncanny valley.

Creating robots and other artefacts that were human enough not to cause unease would, Mori identified, be a major challenge.

As with the Turing Test, it is useful to generalise the uncanny valley phenomenon and apply it to aspects of a virtual human, such as movement or displays of emotion which, in some ways, might be more 'accurate' than those of less well developed systems but which are not realistic enough be to totally taken for a real human and so lie stuck in the 'uncanny valley'.

VIRTUAL HUMANS LANDSCAPE

Figure 2.5 illustrates how virtual humans could be mapped onto the AI landscape presented above. Virtual humanoids, such as customer service chatbots, occupy the middle left of the diagram, their presentation is only partly successful at masquerading as human, and their underlying capability is relatively limited. By contrast, virtual sapiens is very much in the top-right corner, essentially indistinguishable from a human and sentient.

Identifying the exact space occupied by a more general 'virtual human' is more complex as it needs to have enough capability to present as human,

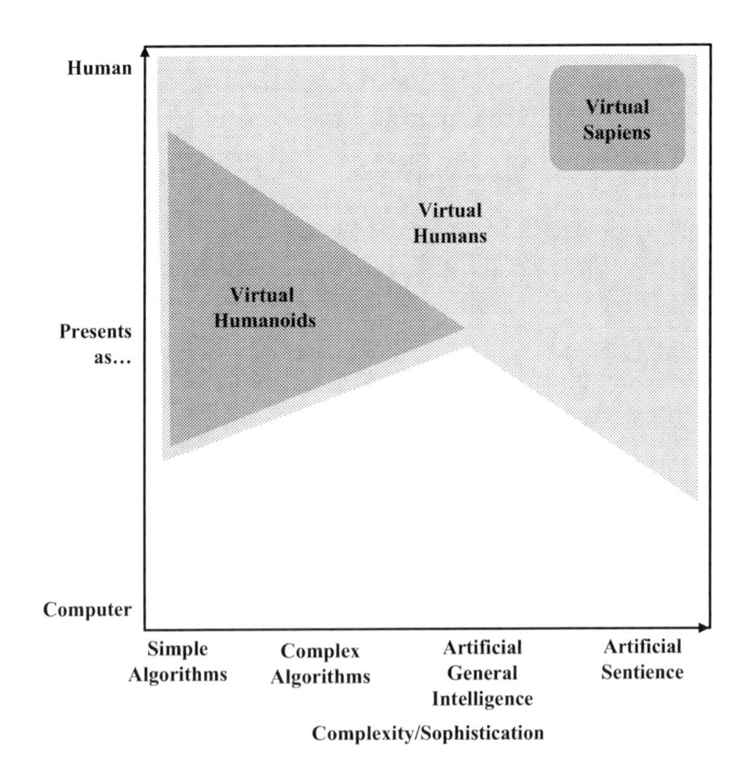

FIGURE 2.5 Virtual humans on the AI landscape.

but there is a trade-off between presentation and the sophistication of its underlying capability. A bot which presented well but had limited capability would be a reasonable virtual human, but so would a bot with more limited presentation but which had the capability of an AGI, and so could act in a more convincingly human (or at least intelligent) way than its appearance might suggest.

ARTIFICIAL INTELLIGENCE VERSUS AUGMENTED INTELLIGENCE

There is often confusion between AI and 'augmented intelligence'. Augmented intelligence is invariably used to describe two different approaches:

- Using a computer (or AI) system that is external to the human to help the human with mental tasks.
- Linking computing capability directly into the brain in order to enhance the brain's processing power or memory.

The second of these is outside the scope of this book. The former is, in many ways, the assumed role of any virtual human anyway by helping humans with tasks, until such time as completely autonomous and self-motivated virtual sapiens and AI come into being.

VIRTUAL HUMANS IN SCIENCE FICTION

It is worth considering how AI, and virtual humans in particular, have been portrayed within science fiction. Whilst science-fiction, with its highly capable and often sentient virtual humans, is no predictor of the future, as Harari (2017) suggests

> If you hear a scenario about the world in 2050 and it sounds like science fiction, it is probably wrong; but if you hear a scenario about the world in 2050 and it does not sound like science fiction, it is certainly wrong.

Film and Television

Alicia Vikander's Ava in Alex Garland's *Ex Machina* (Garland and Macdonald, 2015) is perhaps the most recent direct investigation of AI in film, as Domhnall Gleeson's character acts as a judge in a Turing Test on steroids trying to work out whether an android AI is actually 'conscious', i.e. sentient. Another interesting investigation is in Channel 4's *Humans* (Vincent and Brackley, 2015) with its mix of 'unaware' Synthetic People (Synths) and those which have become aware. Possibly the most interesting character is not Gemma Chan's Mia (the main subject) but Ruth Bradley's D.I. Karen Voss, a sentient Synth fashioned by the Synth creator David Elster to replace his deceased wife. She has lived for many years completely undetected, even holding down a job in the special police unit which is hunting malfunctioning and sentient Synths. Then, in the third series, V is introduced, a virtual human being built by Dr. Athena Morrow and modelled on her dead daughter, and probably a more reasonable representation of what could be built in the next decade.

Except for V, all these examples are very much robotic or android virtual humans. But what of purely digital forms? Malevolent computer programs of the likes of *Terminator*'s Skynet (Cameron and Hurd, 1984), or smart, vocal computers such as *Red Dwarf*'s Holly (Naylor and Bye, 1988), *Iron Man*'s JARVIS (Favreau et al., 2008), or *Blake's Seven*'s Orac (Nation and Maloney, 1978) are seen as common place, and rarely debate their own existence. Slightly more interesting of course is Hal from *2001* (Kubrick and Clarke, 1968), who starts to contemplate his own fallibility and mortality with fatal consequences.

Samantha, the operating system from the Spike Jonze film *Her* (Jonze and Ellison, 2013) is closer to a currently feasible virtual human. There is no attempt at a visual manifestation beyond the voice, but the Samantha character has enough warmth and empathy that the Joaquin Phoenix character falls in love with her. Another example is Robert Picardo's 'Doctor' character, an emergency medical hologram, from *Star Trek: Voyager* (Berman, 1995), although he exists primarily as a projected hologram. *Blade Runner 2049* (Villeneuve et al., 2017) provides two examples of virtual humans. The replicants are virtual human androids, like the Synths in Channel 4's *Humans*. More interesting though, from this book's perspective, is Joi, Ryan Gosling's character K's virtual girlfriend. Initially, she is a disembodied voice in the galley kitchen in his small flat, but then she emerges as a hologram, and shows off an ability to rapidly change outfits to suit his mood. She is obviously highly empathic, and he is certainly in love with her. It also becomes apparent that she is restricted to the flat since she is projected by a device on the ceiling. In one scene, he even hires a prostitute over which he projects Joi's 3D holographic image. Later, K buys a device which lets him project her anywhere he goes and eventually he downloads her whole consciousness to one small device. There are some obvious technical challenges as to how a device in his pocket not only projects her many feet away, but also enables her to take up a viewpoint many feet away from him, but the concept of Joi is otherwise an interesting implementation of a virtual human. It must be noted though that as with *Her*'s Samantha there is certainly a sense of female virtual humans as merely male physical human sex objects (Alexander, 2017).

The *Black Mirror* TV Series has a more interesting take on virtual humans. In the episode entitled 'Be Right Back' (Brooker and Harris, 2013), a woman uses an application on her phone to create a virtual human persona of her recently dead boyfriend, using his social media profile and media recordings. However, when she has the virtual human placed into a physical android body, the experience becomes far too uncanny. Another episode is set in a virtual world which, like *Planet B* discussed below, reflects different historical periods, but is where the dead can upload their consciousness when they die.

Perhaps the best example of a virtual human though is in *Caprica* (Aubuchon and Moore, 2010), the prequel to the reimagined *Battlestar Galactica*. Zoe Graystone, daughter of Daniel Graystone creates a virtual version of herself to inhabit her father's V-World social virtual world (both the world and Daniel are reminiscent of the Second Life social virtual world and its founder Philip Rosedale). Virtual Zoe admits that the technology does not exist to build her from some form of mind-upload of physical Zoe, but that in her life, physical Zoe, like everyone else, had created an extensive digital trail of data records, from email and social media to till receipts, medical records and psychology evaluations that were enough for physical Zoe to use to create virtual Zoe. When physical Zoe dies, the virtual Zoe knows that something is up and rapidly starts to debate her own existence. A similar thing happens when Daniel uses the same technology to create a digital copy of Tamara Adama, who was killed in the same blast as his daughter. The virtual Tamara completely freaks at the absence of any sound of her own heart beating. Virtual Zoe also demonstrates the 'Gleisner' ability to occupy a physical form, taking over a prototype Cylon robot, and maintaining the fiction of the robot being non-sentient as her father tries to get her to kill her pet dog. Virtual Zoe finally escapes back into V-World, but her trace lingers on in the Cylon programming and leads to the sentient Cylons and the rest of the *Battlestar Galactica* story.

Radio

The *Hitchhiker's Guide to the Galaxy* (Adams, 1978) is an early example of AI/robotics in original radio scripts. Whilst it has numerous characters that present in some way as human, from Marvin the Paranoid Android (sentient but extremely cynical) to Happy Vertical People Transporters (prescient and cynical) and Eddy the Shipboard Computer (chirpy and cynical) they are all secondary to the main story.

A less known original radio play is *Planet B* (Broughton et al., 2009). The action again centres around a Second Life type virtual world called Planet B, and in the first series the protagonists are teleported to a different historically themed world every week. Whilst most of the avatars are being controlled by living humans, or are low functioning NPCs controlled by the world, there are also 'rogues', sentient avatars derived from non-human sources, in fact evolved computer viruses. With its Second Life style backdrop, *Planet B*, like *Caprica*, provides an often very believable model of a virtual world occupied by both humans and sentient digital constructs.

Drama

One important example of AI on stage is '*Hello Hi There*' (Garner and Dorsen, 2011), which features two chatbots authored by Loebner prize-winner Robby Garner. The chatbots are on stage as simple computers with projected displays, programmed with material extracted from a famous 1970s debate between the linguist Noam Chomsky and the philosopher Michel Foucault. Alexis Soloski (2011), reviewing the production in the *Guardian*, said

> 'That might not seem the most scintillatingly chatty material, but the computers ran with it. Because the robots can respond to each other with hundreds if not thousands of programmed rejoinders, the text altered enormously each evening. Without facial expression or vocal intonation or much in the way of plot, I'd expected a dull night out, but during the performance I attended, the bots proved unexpectedly charming. They could be amiable, insulting, philosophical, defensive, combative, and sometimes merely random or recursive – in other words, much like the human conversational partners I have known. Even though I knew the computers to be little more than a plastic case and an operating system, I felt engaged with them, especially when they made dumb jokes.'

This is not about weaving a narrative story around the idea of AI, virtual humans and conversational agents, but rather presenting the audience with the technology as it is and leaving them to work out how they relate to it, and what bigger questions it raises.

Siri, produced in 2017 at the Edinburgh Festival Fringe by La Messe Basse with Aurora Nova (Carbonneau and Dauphinais, 2015) is described as

> An actress on stage and iPhone's personal assistant, Siri, as her only partner ... Through a precise game of question and answer, their exchanges expose the bizarre metaphysical dimension of the machine, while blurring the limitations that separate them.
>
> (SUMMERHALLTV, 2017).

There is no attempt to create a fictional AI character, Siri, the virtual assistant, appears as herself, but as with *Hello Hi There* it is still seen as a compelling piece of theatre.

Literature

Science fiction literature is often more effective than film or TV in considering what virtual humans and true artificial intelligence might actually mean. For example, Greg Egan explores AI issues in several of his books. In *Diaspora* (Egan, 1997), much of humanity has transcended to a digital existence living within 'polises' running on supercomputers and downloading into the already-mentioned Gleisner robots when they need to interact with the remaining 'fleshers' – flesh and blood physical humans. In another of his books, *Permutation City* (Egan, 1994), there is again the idea of humans existing as uploaded computer programs, i.e., virtual humans. Of particular note here is the emphasis on how the amount of processing time that a virtual human receives defines how quickly they perceive the world 'happening'. In order to stow away on a spacecraft, the protagonist's virtual humans steal only a few processing cycles every second, so for them the whole journey apparently happens very quickly since they are running so slowly. There is even an exploration as to whether the AIs could run on a physical processing substrate, something like an abacus, which, if they are just an algorithmic program, should be perfectly possible.

A key figure in cyberpunk literature is Neal Stephenson. His *Snow Crash* novel (Stephenson, 1992), apart from being the inspiration for virtual worlds such as Second Life, features a virtual human called The Librarian who helps Hiro Protagonist access information, but is more sentient AI than Siri. Hiro suspects '... that the Librarian may be pulling his leg, playing him for a fool. But he knows that the Librarian, however convincingly rendered he may be, is just a piece of software and cannot actually do such things' (Stephenson, 1992).

Alastair Reynold's 2013 novel *On the Steel Breeze* (Reynolds, 2013) features one interesting virtual human character – Eunice. In the earlier novel *Blue Remembered Earth* (Reynolds, 2012), Eunice had been the aging matriarch of a powerful family. In the sequel, Eunice is a virtual persona put together from fragments of information about Eunice to help the protagonists with their current quest. Eunice is presented with realistic gaps in her knowledge, but from a synthesis of what she does know about her progenitor is able to help try and fill in the gaps. In the final book of the trilogy, *Poseidon's Wake* (Reynolds, 2015), Eunice has become both organic and sentient with the help of a race called the Watchkeepers, but she is able to recall how she started off as a simple software, which, like Virtual Zoe, was created from public and private records of herself (Reynolds, 2015).

Another interesting exploration of virtual humans is in Tony Ballantyne's 2004–2007 *Recursion* trilogy – *Recursion*, *Capacity* and *Divergence* (Ballantyne, 2004, 2005, 2007) – which features several AI and 'personality constructs'. Whilst *Capacity* features such personality constructs experiencing groundhog-day-type repetitive loops of existence from which they try to find a way out, it is the Kevin AI character from the final book, *Divergence*, which is one of the most interesting. Kevin admits that he is just software running on a processing network, denying that he is an 'AI' and reaffirming that he is just a set of binary logic decisions (Ballantyne, 2007). This is despite one of the other protagonists claiming that Kevin passes the Turing Test every day (Ballantyne, 2007).

That is perhaps one of the key questions of this book: To what extent can a virtual human be created which is convincing, but which is effectively just a set of binary processing decisions, just a massive, if complex, algorithm?

Games

A final source of science fiction views on AI are games. Whilst many videogames present a quite one-dimensional 'evil-AI', or feature the 'AI-as-assistant', as with Cortana (Master Chief's constant virtual mentor 3D hologram) in the *Halo* games (Bungie, 2001), there are a few interesting examples from both computer and pen-and-paper roleplaying games (RPGs) which merit fuller consideration.

In the former category is *Façade* (Mateas and Stern, 2005), a relatively lo-fi game visually where the 'player' sits in on a couple having an argument. How the player then interacts with each of them has a direct bearing on how the whole story plays out. The game was created by Andrew Stern and Michael Mateas, and the chatbot system driving it has 2000 dialog pairs, but a player would only experience about 20% of them on any single play-through of the game.

The roleplaying game (RPG) treatment of virtual humans is particularly relevant as they present a richer, more complete exploration of the future than almost any other form as the very nature of the roleplaying milieu is to live and operate in that future, rather than just be a passive observer or undertake a specific, and often heroic, role.

David Pulver's *Transhuman Space* (Pulver, 2002), published by Steve Jackson Games is a near-future, cyberpunk style game, set within the Solar System of 2100. Even though it was written in 2002, many of its ideas in both space and digital technology still seem quite insightful. In particular, the setting has a well-developed taxonomy of the different sort of AI that live on the computer networks. Any form of 'living' (i.e., sentient or near-sentient) software code is called an infomorph, and may exist either only on the net or occupy a physical body (robot, android or wetware). Two different types of infomorph are identified, Mind Emulations and digital native AIs.

The Mind Emulations are programs that were once organic (human) intelligences but were copied or uploaded into a digital form. This class is further broken down into:

- Ghosts – A perfect digital upload of a being from the organic source (i.e,. brain/mind).
- Shadows – A low-resolution copy of a still living being's mind.
- Fragments – A failed upload that's missing some or all of its memories.

Conversely, the native digital intelligences are split into:

- SAI – Sapient Artificial Intelligence. Human-level intelligence and emotions, all but impossible to tell from a regular human.
- LAI – Low-sapient Artificial Intelligence. Intelligent, but has difficulty with emotions.
- NAI – Non-sapient Artificial Intelligence. Non-intelligent, unemotional, primarily just a tool.

Between the two sit eidolons – manufactured AIs that mimic a real person – and which include memorial-AIs of dead relations (see Chapter 12). Pulver also discusses Emergent Intelligences (EIs), AIs which have attained sentience rather than being created with it, Orphans, an AI which was originally set relatively general goals by its owners, but the owners are now either dead or have neglected it and so it continues on what it thinks is the best course to fulfil those goals, and Rogues. Rogues are EIs or Orphans which have taken a belligerent stance against humanity, and the risks of EIs going rogue, and/or of creating their own cloned or child AIs without any human control haunt the book's Transhuman society. Also illegal is the process of XOXing, creating multiple copies of the same personality (i.e., multiple ghosts), although multiple lower resolution shadows may be permitted.

This taxonomy may have its origins in a science fiction roleplaying game, but it would seem a useful means of analysing the types of AI which are likely to be of interest, and prompt consideration of ways in which virtual humans and AI may interact and co-exist in the future.

Another science fiction roleplaying game, which has a well-developed approach to both virtual humans and virtual worlds, is Mindjammer (Newton and Snead, 2016). Although set several millennia into the future, the game is more focused on exploring the 'soft' transhuman issues of the twenty-first century than the spaceship-and-blaster space opera of twentieth century science fiction. It feels more like *Farscape* than *Star Wars*.

The Mindscape is the central element of Mindjammer. It is a universal internet which provides everyone with instant access to information, systems and each other. Most people have direct brain interfaces to the Mindscape. Newton describes several interesting features of the Mindscape and how it is used:

- People create (and adventure in) virtual environments within the Mindscape (like a futuristic version of Second Life).
- A person's 'halo' is the extent to which they use the Mindscape to augment their everyday life, e.g., using 'exomemories' to both store memory and to acquire new skills and information. These halos can have a public version, like a Facebook page, which others can access and which can 'handshake' information with other people.
- A person can make an explicit 'thought-cast' to another individual – effectively direct brain-to-brain communication.

- A person can take actions at a distance through the Mindscape (e.g., by accessing a distant computer system, robot or even another individual – something known as 'technopsi' – and with observable results which may seem little different from psionic powers or magic.
- The Mindscape contains Memoplexes, collections of exomemories that are shared by groups of individuals. This isn't just a passive set of Facebook Group pages, but the ability for every member to share the same skills, knowledge and collective memories.

One interesting aspect of the Mindscape is that it is limited to the speed of light and conventional communications. As a result, the Mindscape is actually a series of Mindscape Instances in each system or on each world (or even each Starship), each having local information in detail but only a fuzzy or older version of other information. Starships (Mindjammers) and communications links try and keep these instances in sync, but conflicts between updates can cause 'chronodisplacement' when you have two conflicting exomemories!

With specific regards to virtual humans, the Mindscape is also home to:

- Fictionals – exomemories which may be linked to automated Mindscape entities that simulate people and memories that either don't currently exist, or never existed – Virtual NPCs for instance.
- Sentinels – non-sentient AIs which police the Mindscape and, in particular, identify undisclosed Fictionals.
- Eidolons – synthetic, usually digital, artificial life-forms, from intelligent starships to Virtual Persons and Personas. Eidolons must meet the Eidolon Compliances: integrity, stability, freedom from neurosis and freedom from psychosis. Eidolons which fail the Compliances may suffer an Eidolon Crisis with various detrimental results.
- Thanograms – a personality and memory snapshot at the time of someone's death, which may be stored as exomemories for ad-hoc access, or be instantiated as an eidolon – so creating a Digital Immortal. Eidolon know, though, that they are separate to the subject of the Thanogram and may even revere or worship them.

Virtual NPCs, Persons, Personas and Digital Immortals will all be discussed in more detail in Chapter 8.

So, perhaps, one of the best ways to understand what a future that involves virtual humans may be like is to go out and play one of these roleplaying games. You could even play the virtual human!

CONCLUSION

Figure 2.6 maps of some of the fictional AI, virtual humans and robots just discussed onto the AI landscape chart. Physical robots/androids are shown in italics. Science fiction may not provide much useful information about how to build a virtual human, but it provides ample perspective on what it might be like to interact, live and work with one, and on what their own view of themselves might be. Often, the virtual human, sentient robot, android, or AI is just a peripheral part of the story, or a powerful nemesis for the lead characters. It is those stories where the virtual human is at the centre of the tale, and often struggling with its own sense of what it is to be a virtual human, which can provide a useful insight into some of the ethical, moral and philosophical issues surrounding virtual humans.

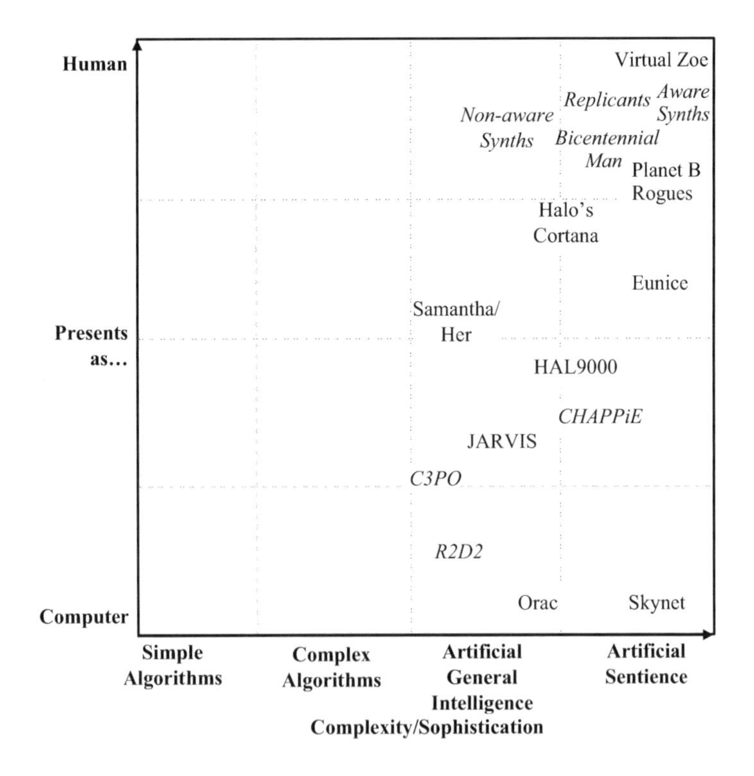

FIGURE 2.6 Mapping fictional AI and virtual humans.

Having established the scope of what a virtual human is, how it relates to terms such as artificial intelligence, and what the science fiction visions of virtual humans are, Part II of this book will look at the various technologies which contribute to the creation of a virtual human, and make an assessment of how far advanced each of these technologies are towards creating a sophisticated virtual human worthy of the science fiction vision.

REFERENCES

Adams, D. (1978). Hitchhiker's Guide to the Galaxy [Radio series]. London, UK: BBC.

Alexander, J. (2017). Blade Runner 2049 continues questionable trend of the 'algorithm-defined fantasy girl'. *Polygon.* Available online https://www.polygon.com/2017/10/11/16455282/blade-runner-2049-analysis-ana-de-armas-fantasy-girl.

Aubuchon, R., & Moore, R. D. (Creators). (2010). Caprica [TV Series]. New York: Syfy.

Ballantyne, T. (2004). *Recursion.* London, UK: Macmillan.

Ballantyne, T. (2005). *Capacity.* London, UK: Macmillan.

Ballantyne, T. (2007). *Divergence.* London, UK: Macmillan.

Berman, R. (1995). Star Trek: Voyager [TV series]. Los Angeles, CA: CBS Television.

Bradeško, L., & Mladenić, D. (2012). A survey of chatbot systems through a Loebner prize competition. In *Proceedings of Slovenian Language Technologies Society Eighth Conference of Language Technologies* (pp. 34–37). Ljubljana, Slovenia: Institut Jožef Stefan.

Brooker, C., & Harris, O. (2013). Be Right Back. Black Mirror [TV Episode]. Westminster, UK: Channel 4.

Broughton, M. et al. (Creators), & Dromgoole, J. (Producer). (2009). Planet B [Radio series]. London, UK: BBC.

Bungie. (2001). Halo [Computer Game]. Bellevue, WA: Microsoft.

Cameron, J. (Writer & Director), & Hurd, G. (Producer). (1984). The Terminator [Motion picture]. Los Angeles, CA: Orion Pictures.

Carbonneau, M. (Director & Creator), & Dauphinais, L. (Creator). (2015). Siri [Play]. Canada.

Egan, G. (1994). *Permutation City.* London, UK: Orion.

Egan, G. (1997). *Diaspora.* London, UK: Orion.

Epstein, R., Roberts, G., & Beber, G. (Eds.). (2009). *Parsing the Turing Test.* Dordrecht, the Netherlands: Springer.

Favreau, J. (Director), & Fergus, M. et al. (Screenplay), & Arad, A. (Producer). (2008). Iron Man [Motion Picture]. Los Angeles, CA: Paramount Pictures.

Garland, A. (Writer & Director), & Macdonald, A. (Producer). (2015). Ex Machina [Motion picture]. London, UK: Universal Pictures.

Garner, R. (Writer), & Dorsen, A. (Director). (2011). Hello, Hi There [Play]. United States.

Gibson, W. (1996). *Idoru.* New York: Viking Press.

Harari, Y. N. (2017). Life 3.0 by Max Tegmark review – we are ignoring the AI apocalypse. *The Guardian.* Available online https://www.theguardian.com/books/2017/sep/22/life-30-max-tegmark-review.

Hendler, J. A. (2008). Avoiding another AI winter. IEEE Intelligent Systems, Vol. 23, No. 2, 2–4.

Jonze, S. (Director, Writer & Producer), & Ellison, M. (Producer). (2013). Her [Motion picture]. Burbank, CA: Warner Bros.

Kubrick, S. (Director & Producer & Screenplay), & Clarke, A. C. (Screenplay). (1968). 2001 [Motion picture]. Los Angeles, CA: Metro-Goldwyn-Mayer.

Lea, A. (2017). Beware Fake AI! Linked-IN web site. Available online https://www.linkedin.com/pulse/beware-fake-ai-andrew-lea.

Mateas, M., & Stern, A. (Designers). (2005). Façade [Computer game]. Pittsburgh, PA.

Mori, M. (1970). The uncanny valley. *Energy, 7*(4), 33–35.

Nation, T. (Creator), & Maloney, D (Producer). (1978). Blakes Seven [TV series]. London, UK: BBC1.

Naylor, G. (Creators), & Bye, E. (Director & Producer). (1988). Red Dwarf [TV series]. Birmingham, UK: BBC 2.

Newton, S., & Snead, J. (2016). Mindjammer: Transhuman Adventure in the Second Age of Space [Role-playing game]. Harlow, UK: Mindjammer Press Ltd.

Pennachin, C., & Goertzel, B. (2007). Contemporary approaches to artificial general intelligence. In B. Goertzel & C. Pennachin (Eds.), *Artificial General Intelligence* (Vol. 2). New York: Springer.

Powers, D. M. (1998). The total Turing test and the Loebner prize. In *Proceedings of the Joint Conferences on New Methods in Language Processing and Computational Natural Language Learning* (pp. 279–280). Stroudsburg, PA: Association for Computational Linguistics.

Pulver, D. L. (Designer). (2002). Transhuman Space [Role-playing game]. Austin, TX: Steve Jackson Games.

Reynolds, A. (2012). *Blue Remembered Earth.* London, UK: Gollancz.

Reynolds, A. (2013). *On the Steel Breeze.* London, UK: Gollancz.

Reynolds, A. (2015). *Poseidon's Wake.* London, UK: Gollancz.

Saygin, A. P., Cicekli, I., & Akman, V. (2000). Turing test: 50 years later. *Minds and Machines, 10*(4), 463–518.

Soloski, A. (2011, 28 January). No actors. Just robots. Call this a play? *The Guardian.* Available online https://www.theguardian.com/stage/theatreblog/2011/jan/28/no-actors-robots-play-theatre.

Stephenson, N. (1992). *Snow Crash.* New York: Bantam Books.

SummerhallTV. (2017). Laurence Dauphinais & Maxime Carbonneau: SIRI. *SummerhallTV.* Available online http://www.summerhall.tv/2017/laurence-dauphinais-maxime-carbonneau-siri/.

Turing, A. M. (1950). Computing machinery and intelligence. *Mind, 59*(236), 433–460.

Villeneuve, D. (Director), Fancher, H. (Screenplay), Kosove, A. et al. (Producers). (2017). Blade Runner 2049 [Motion Picture]. Burbank, CA: Warner Bros.

Vincent, S., & Brackley, J. (Creators & Producers). (2015). Humans [TV series]. Westminster, UK: Channel 4.

II

Technology

INTRODUCTION

Part II presents the various technologies used to create a virtual human. As explained in Chapter 1, the concern in this book is just with digital virtual humans, so robot-related electro-mechanical (and other) technology will not be considered. This book also focuses on what can be done with the technology in early 2020s, rather than dwelling on the aspirational technologies of the future – although some of these will be considered in Chapter 13. However, consideration will be given as to what factors and interests are prompting the development of the different technologies, and how rapidly progress might be made. Particular challenges that the development of a contributing technology introduces, such as ethical or moral concerns, will also be explored.

In order to describe the various elements of a virtual human, the component model in Figure II.1 will be used. It is a more detailed version of the model introduced at Figure 1.1. This has been designed to act as a guide to the discussions in the following chapters. There is no clear agreement or model of what constitutes a virtual human, but the elements shown in Figure II.1 are largely those that appear with regularity across different researchers' cognitive and virtual human architectures as described in more detail in Chapter 6, and so can be taken as being a reasonable

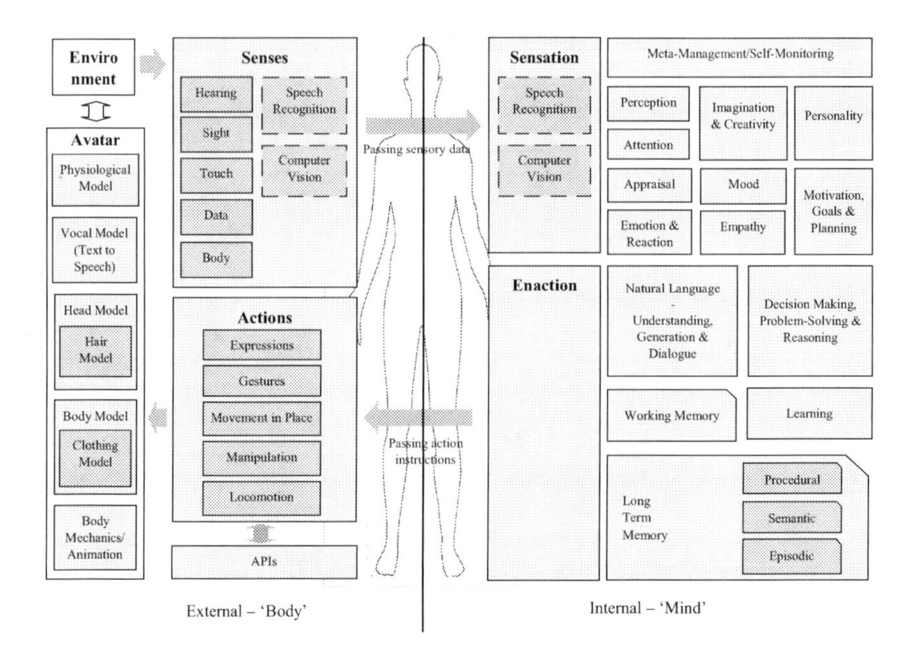

FIGURE II.1 Virtual human component architecture.

representation of the key elements of a virtual human. This model has also been found, in the authors' own work, to be at a practical level to give a useful level of detail without becoming too complex.

It should be noted that the model shows a clear separation between the 'body model' and the 'mind model'. Maintaining this distinction emphasises the fact that the virtual human's 'body' could exist in a wide variety of manifestations, from disembodied voice on a mobile phone, to a 3D avatar in a virtual world, to even a physical robot, possibly all at the same time. In practice, this book argues that such a distinction between the body and mind elements may not be as unambiguous as it is initially presented here, but for now, it is a useful distinction. It should also be noted that in this book 'mind' and 'brain' tend to be used synonymously, whereas a stricter definition might be that, for Chapter 4 in particular, the 'mind' represents the processes whereas the 'brain' represents the processing substrate, although such dualism is, of course, open to debate!

Each element of the model will be discussed in more detail in the chapters that follow.

- Chapter 3 will consider the elements of 'body' and the senses.
- Chapter 4 will consider the elements of the 'mind'.

- Chapter 5 will consider communication.
- Chapter 6 will consider the more functionally orientated cognitive architectures.
- Chapter 7 will consider the environment within which the virtual human may operate and the importance of embodiment.
- Chapter 8 will then consider how these different elements can be brought together to create a virtual human, and how 'human' such a virtual human might be.

CHAPTER 3

Body and Senses

INTRODUCTION

This chapter will examine the technologies which enable the creation of an avatar body for a virtual human, and the ways in which that body can incorporate human, and non-human like senses. It will explore the extent to which current approaches and techniques enable a human body to be modelled realistically as a digital avatar and analyse how the capability might develop over the coming years. The section headings follow the elements shown in the Virtual Human Component Architecture at Figure II.1.

Much of the work in this area is being driven by the computer generated imagery (CGI) of the film industry, and the motion capture and animation of the gaming industry, where CGI is almost indistinguishable from the real. However, for a virtual human, both imagery and animation need to be generated in real time, and in response to unknown events. This makes it more challenging than movie CGI, but mirrors the trend in many computer games towards having 'sandbox' environments where the player can explore well beyond a scripted set of encounters.

WHAT MAKES AN AVATAR?

In digital terms, an avatar for a virtual human (or physical human) can take a number of forms, including:

- A static 2D head-and-shoulders image, as used in many chat applications,
- An animated 2D head-and-shoulders or full body image, as used in some customer support applications, or
- A fully animated 3D character within a game or virtual world.

In creating an avatar for a human or virtual human, the key technology areas that need to be considered are:

- Facial rendering,
- Facial and speech animation,
- Hair modelling,
- Body rendering and modelling,
- Body mechanics and animation,
- Clothes modelling, and
- The body physiological model.

These are shown diagrammatically in Figure 3.1.

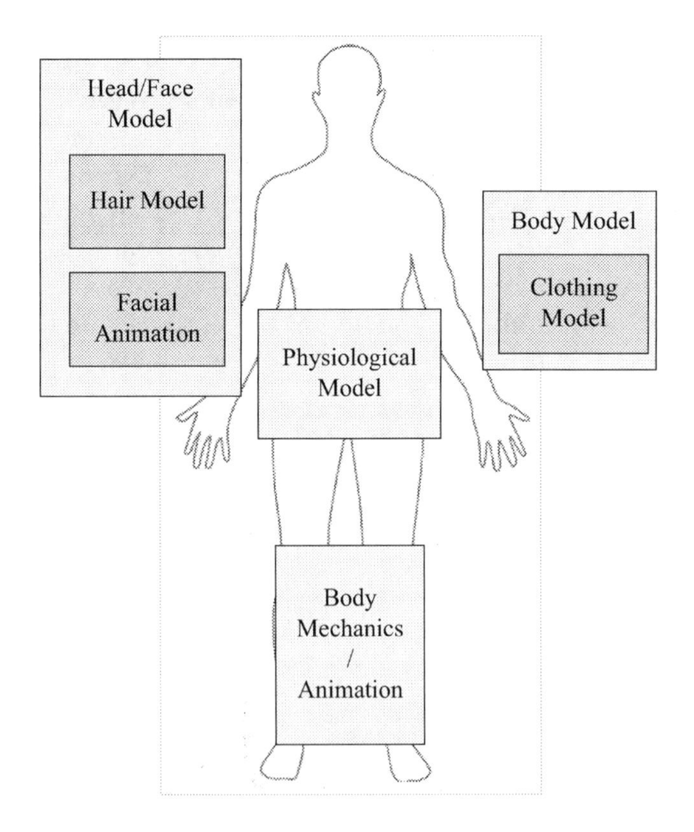

FIGURE 3.1 The elements of a virtual human body.

Note: Colour images and video of state-of-the-art examples to illustrate each of the areas discussed, along with further links, are on the website accompanying this book at www.virtualhumans.ai.

FACIAL RENDERING

Probably the most important element of any avatar representation of a virtual human is the face (Figure 3.2). For humans, talking face-to-face with someone else is seen as a richer experience than talking on the phone using only voice, or having a text-chat. There is also a preference for a physical face-to-face rather than a video link, partly since the higher fidelity enables more subtle conversational and emotional cues to be detected.

The challenge of accurate facial rendering is dominated by the notion of Mori's 'uncanny valley' (1970) described in Chapter 2.

In one experiment, Fan et al. (2012) showed facial parts, in real and CGI pairs, to participants and identified that the main clues that enabled people to tell a human from a CGI face were: Eyes (36%), Skin (22%), Illumination (17%), Expression (11%) and Colour (2%). In addition, research by MacDorman et al. (2009), whilst confirming much of the uncanny valley hypothesis, also found that it was vital to get facial proportions right, especially as the photorealism increases. In a further study, Farid and Bravo (2012) found that people could identify the 'real' image of a person 85% of the time with typical (2010 era) computer generated faces, i.e., a computer generated version only fooled them 15% of the time. Whereas Balas and Pacella (2015) found that artificial faces formed an 'out-group', that is, they were seen as qualitatively different from the faces that a person usually sees (for example, in terms of race or

(a) (b) (c)

FIGURE 3.2 Simple 2D avatars developed with the Sitepal system (www.sitepal.com). Animated, talking versions are available at www.virtualhumans.ai, as well as links to more state-of-the-art facial renders. (a) A cartoon 2D animated avatar. (b) An animated 2D avatar based on a photograph of one of the authors. (c) An animated 2D avatar based on a photograph of one of the authors.

age) and so were harder to remember or discriminate between than real human faces. From a technical modelling perspective, the key elements of a high-quality face are:

- A mesh (3D model) with sufficient detail (and deformation control) to mimic even the smallest creases in the face and provide believable movement (see next section);
- 'Normal' maps to show very minor skin detail, e.g., pimples, pores;
- Texture imagery with constant variation across the face to provide believable skin tone and blemishes; and
- Specular maps and other approaches (for example, sub-surface scattering) so that the skin responds correctly to light illumination and doesn't look too waxy (based on Staples 2012).

Many commercial simulation applications, such as Virtual Battlefield System 3, use 'photo-faces', where a conventional 2D face image is 'pasted' onto a generic or customizable head shape. The face can be mapped to the head shape based on features such as the eye-eye distance. Compared to the CGI offerings, the faces can look very anomalous, particularly when animated or viewed from side angles, and can easily fall into the uncanny valley. Many consumer games have adopted more cartoon-like avatars, possibly until such time as the more lifelike faces are available.

The real challenges in facial animation are:

- Being able to render in real time, particularly in response to movement and changes in lighting;
- Rendering hair; it is no coincidence that many examples of high quality facial rendering tend to have no hair or very close cropped hair; and
- Being able to capture the likeness of a real person in an automated way.

The key limitation appears to be computer processing power, given the amount of light modelling and mesh detail required for highly realistic images and animations. This means that mobile devices are likely to have inferior avatar faces for the medium term.

FACIAL AND SPEECH ANIMATION

Whilst having a high-fidelity digital model of a human face offers one level of problem, having it move to create realistic expressions, and, particularly, to synchronize its mouth with any speech is an even greater challenge. Facial animation includes both the making of facial expressions (raising eyebrows, scrunching eyes, and smiling) and the movement of the mouth (and neighbouring areas) to match any speech being produced. A text-to-speech engine (see Chapter 5) generates the sound of what the avatar is saying, but this sound must then be used to animate the lips in real time so that it seems that it is the avatar that is saying it.

Kshirsagar et al. (2004) describes the lip sync process as being:

1. Extracting the phonemes from the speech audio. Phenomes are the smallest units of speech sound, such as 'p', 't', 'th', 'ng' and there are about 46 phonemes in English.

2. Mapping phonemes to visemes, the visual counterparts of each phoneme showing lip shapes, such as pursed lips for an 'oo' sound, from a parameterized database.

3. Blending between successive visemes (called co-articulation).

The broader facial expressions must of course also be consistent with the spoken text (for example, eyebrows, eyes, head movements).

Ezzat et al. (2002) describes three approaches to speech animation:

1. Keyframing – the viseme method described above. For example, Ezzat describes how speech animation can be applied to video output, so that from one reference video, the videoed person can be made to say anything.

2. Physics-based, modelling the underlying facial muscles and skin. Nagano's work (2015) follows this approach, using a microstructure approach to skin deformation during talking and other facial gestures. A video of the system is available at http://gl.ict.usc.edu/Research/SkinStretch/.

3. Machine learning methods trained from recorded data and then used to synthesize new motion (for example, hidden Markov models – HMMs). This approach is used by Cao et al. (2005) who extends the machine learning control of speech animation to include all related

emotion driven facial expressions as part of the speech. It is also used by Cassell et al. (2004) in the Behaviour Expression Animation Toolkit (BEAT), developed at the Massachusetts Institute of Technology (MIT) with Department of Defense (DOD) funding to automatically and appropriately fully animate the face from the text that it is meant to say.

Other video demonstrations of speech animation can be found at www.virtualhumans.ai.

Speech is a far more natural way of interacting with virtual characters, but if the sound is not linked to the face of the character speaking, then it appears disembodied. However, once linked to a face, the speech animation/lip sync must be good enough not to be off-putting and, as a result, cause an interruption to the illusion. Thus, better speech animation should result in improved simulations and virtual character engagement, particularly where interpersonal skills are required. However, the ability to create a fully lifelike, talking head of another person opens up a wide variety of ethical issues.

HAIR MODELLING

It is notable that many of the published examples of high-fidelity facial models are either bald or wearing hats. This is because the realistic modelling of hair presents its own set of unique challenges. As with the face, whilst modelling static hair offers significant problems, these are amplified when that hair needs to move in response to the movement of the head, of the character (for example, when moving quickly or running a hand through the hair), or to wind. As a result, the modelling of human hair is seen as a significant challenge and quite different from that of modelling the face and the rest of the head. Hadap and Magnenat-Thalmann (2001) identify three main tasks in hair simulation: hair shape modelling, hair dynamics and hair rendering.

Hair Shape Modelling

This attends to the geometry of the hair, and the density, distribution, and orientation of the different hair strands. Ward et al. (2007) describes hair shape modelling as comprising three main steps: attaching hair to the scalp, giving the hair an overall shape, and managing finer properties of the hair. There are a variety of ways to define the overall shape, including parameterized geometry, physical-based models, and imagery of real hairstyles, and the most effective systems seem likely to build on the strengths of different methods.

Hair Dynamics

Hair dynamics describes how the hair moves in response to movement of the body, wind or other factors. Early models used a mass-spring approach (Rosenblum et al. 1991) and Bertails et al. (2006) discusses more recent approaches using two dimensional projective dynamics, chains of rigid bodies and Kirchhoff models.

Hair Rendering

Hair rendering involves giving colour to the hair and needs to take into account the complex light and shadow effects caused by the layers of hair, and the semi-transparency of some hair strands themselves. Daldegan et al. (1993), as part of a description of an integrated system for modelling, animating and rendering hair, used a ray-tracing approach, as is used in many other high-fidelity rendering tasks. Leblanc et al. (1991) describes a pixel blending approach and Sintorn and Assarsson (2008) uses opacity shadow maps to deal with the problems of self-shadowing (one strand casting a shadow on another) and transparency.

As well as the hair modelling and simulation itself, there are also challenges in modelling different ethnic types of hair (Patrick et al., 2004).

Without good simulated hair, the high-quality rendering and animation of an avatar can be worthless as any sense of true believability of the virtual human can be lost.

BODY RENDERING/MODELLING

Whilst head and shoulder avatars are adequate for many virtual human uses, such as virtual customer service assistants on web sites or virtual personas in artificial Skype conversations, the deployment of virtual humans into virtual worlds or games requires them to acquire a full body avatar.

Realistic rendering of the human body presents many of the same challenges as rendering the human face, although it can be more forgiving, and is often covered by clothing (which is a separate challenge in itself). The initial concern in this section is in what might be described as ‘the body at rest’. The following section will consider the animation required to create a moving body. A first requirement is to understand the dimensions of a human body. Seo (2004) describes the research in this area of anthropometry, measuring the human body. This work has become significantly easier with the advent of 3D cameras to capture depth information (for example, http://www.bodysnapapp.com/). The proposed H-Anim standard defines 75 key measurement points on the human body for avatar modelling and the MPEG4 BPA standard (Visage, 2012) defines 196 Body Animation Parameters.

Adabala (2004) describes four main features of skin rendering, including colour and micro-geometry, illumination, interaction with the environment and temporal features (including wrinkling and pallor), with similar issues and approaches to those found in facial modelling. Commercial virtual worlds, simulations and game systems include body rendering systems of different levels of sophistication. Stuart (2015) provides a useful summary of Triple-A game avatar realism (as well as scenery and emotional gameplay). The most detailed body renders are to be found in film CGI where the use of non-real time rendering enables more or less completely realistic bodies to be created.

As well as obvious applications within training simulations, a further interesting area for commercial use is in fashion and the idea of creating an avatar of oneself in order to try on clothes (Brooke, 2014). This is one of the few areas where the focus is on creating avatars of specific real people, rather than of fictional characters.

BODY ANIMATION/MECHANICS

As with the face and hair, animating the body introduces a whole new set of challenges. Many good-looking full body models can become ugly and unrealistic if not correctly animated. Whilst some roles may require a relatively limited set of animations (for example, a virtual tutor showing engine repair or a soldier NPC in a first-person shooter), the ultimate aim must be to enable the virtual human to perform any human animation. Interestingly though, the physical humans in virtual worlds usually only have access to the same animations as the virtual humans, and it is easier for the virtual humans to trigger those animations, so the virtual humans can seem more human than their physical counterparts.

Body animation is often accomplished by rigging (attaching) the 3D body model 'mesh' to a hidden 'skeleton', made up of software 'bones' and 'joints', which are then animated by code or user action, causing the attached mesh to bend and deform accordingly.

Boulic et al. (1990) provides an introduction to body motion control and highlights the two main approaches: kinematic, the geometry of motion regardless of its physical realization, and dynamic, based on Newton's laws of motion, relating to forces and mass. Most systems take a kinematic approach, where the aim is to ensure that the animation is visually realistic. Inverse Kinematics (IK) (Tolani et al., 2000) is one of the most commonly used approaches.

Major challenges in body animation include how to handle deformation of the body model/mesh at joints which have a lot of travel so that the body doesn't take on an unnatural shape (Yang et al., 2015). Possible solutions include the Dyna model (Pons-Moll et al., 2015), which can take into account different Body Mass Indexes, and the 'Skinned Multi-Person Linear (SMPL) model of the human body that can represent a wide range of human body shapes realistically, can use natural pose-dependent deformations, exhibits soft-tissue dynamics, is efficient to animate, and is compatible with existing rendering engines' (Loper et al., 2015).

Given its dynamic nature, body animation is best seen through video, and examples can be found at a variety of locations on the Internet and at www.virtualhumans.ai.

One of the biggest challenges in body animation is in the standardising of the human reference models and animation files so that they can be shared between simulation systems, and even between characters. Kinematic systems can also result in significant deformations of the body mesh when more extreme, or poorly mapped, movements are represented. Whilst the representation of individual body animations is now very realistic, the challenge is how to string them together into long fluid sequences in response to real time triggers, whether from humans or virtual humans.

CLOTHES AND CLOTH MODELLING

Whilst there are undoubtedly some, perhaps dubious, applications of virtual humans which would be content with them remaining naked, most situations are going to require them to be clothed. The modelling of clothes requires similar activities to the modelling of hair. The geometry of the clothing (its shape and style) needs to be defined, as well as how it moves in response to the movement of the body. However, whereas hair modelling is dealing with thousands of elements whose individual textures are relatively unimportant, it is the effect of the whole which is of interest, with clothing, only a handful of items are being modelled, but the accurate modelling of their texture is vital.

Adabala (2004) describes the key features of clothes rendering at a very fine detail, including model thread illumination, knit/wave milli-geometry, and illumination, as well as the macro-geometric movement (how it folds, flows and wrinkles). Frâncu (2015) presents a variety of methods used to model the structure and dynamics of clothing. As with modelling methods, there is a trade-off between accurate but slow (mechanical methods) and fast but inaccurate (geometrical methods). Sattler et al. (2003) describes how the mesostructure of cloth 'is responsible for fine-scale shadows, occlusions, specularities and subsurface scattering effects. Altogether these effects are responsible for the "look and feel" of cloth' (Sattler et al., 2003: 167). Daubert et al. (2001) describes a similar but more efficient approach. Some researchers are even going down to yarn-level simulations, such as Wu and Yuksel (2017) and Cirio et al. (2014). The collision of clothing items with the body, each other, and other objects, is also important (Volino and Cordier, 2004).

Aliaga conducted a set of experiments to explore which factors contributed to the perception of cloth and to determine how efficiency of the modelling could be improved without sacrificing the perceived realism. Aliaga found that 'appearance dominates over dynamics, except for the few cases where dynamics are very characteristic, such as in the case of silk'. (Aliaga et al., 2015). In addition to the demands from games and film CGI driving clothing modelling research, there is also increasing interest in cloth modelling in the fashion industry for both computer-aided garment design and to allow shoppers to try on fashion items virtually at home (called 'Virtual Try On' – VTO) before buying (Zhang et al., 2015).

A different challenge for virtual clothing is provided by Miguel who notes that clothing manipulation (for example, getting dressed) 'is a common action in humans that many animated characters in interactive simulations are not able to perform due to its complexity' (Miguel et al., 2014). Miguel identifies the steps required to perform a getting dressed task and how the animation of that can be undertaken and how it has an impact on the virtual cloth, taking into account all the clothing collisions and deformations, and recommends changes to cloth and character animation models to improve the avatar capabilities.

BODY PHYSIOLOGICAL MODEL

Whilst it is understandable that a virtual human would need a visual avatar when interacting with physical humans, it is less clear why they may also need to model how the internal organs and systems of their non-existent physical body are working. However, there are benefits in some use cases to the virtual human having such a model.

This type of model is referred to as a physiological model. When used with an avatar or virtual human, a physiological model can be used to enable an avatar to replicate the performance or responses of a physical human. For instance, this may define the interactions between different parameters, such as walk and run speeds, load carrying ability, stamina, level of dexterity, and the impact on these of health and environmental conditions, such as heat and cold. The physiological model can also include trauma models, which reflect changes in performance due to injury, and which may model the avatars 'dying', and pharmacological models, which model the impact of drugs, and possibly other treatments.

Much of the research and literature is based on either medical physiological models (concerned with treatment of 'virtual patients'), or on the health applications of physiological models. The games sector does not appear to be interested in performance constraining physiological models (indeed, most avatars are superhuman, to an extent, to encourage gameplay). In the health arena, Díaz-Zuccarini et al. (2014) describes the work on the €72 million European Commission-funded European Virtual Physiological Human (VPH) Initiative (https://ec.europa.eu/digital-agenda/en/virtual-physiological-human). As UCL (2015) reports

> 'Models of … parts of the body describing kidney function, joint and limb mechanics, and blood flow (amongst others) all exist. The challenge faced by researchers is how to connect these models to create a more effective and realistic simulation or the human body, both in health and disease. It is the goal of the VPH initiative to create the research environment in which progress in the investigation of the human body as a single complex system may be achieved.'

A Virtual Physiological Human Institute has been set up to support the work. (http://www.vph-institute.org/)

Kofránek et al. (2013) discusses a more practical virtual physiological model called HumMod (http://hummod.org), which describes itself as 'The best, most complete, mathematical model of human physiology ever created' with 5000 model variables and implemented in Extensible Markup Language (XML). Kramar et al. (2013) provides a wider survey of current Virtual Physiological Human initiatives.

The way in which physical human performance can be linked to avatar performance is described by Bartley et al. (2013) giving the example of a

> 'mobile role-playing game (RPG) where the character evolves based on the exercises the user performs in reality … This novel application has shown the capability of automatically identifying and counting the exercises performed by the user … The type and amount of exercise improve the characters speed, strength, and stamina based on the type and amount of exercise performed.'

In a virtual human, it could be that virtual exercise is needed to enhance the virtual human's avatar ability – so grounding the experience of fitness. A key issue is the difference in levels of fidelity that different use cases demand of the physiological model. The medical fraternity is understandably seeking a very high-resolution model, as it will drive life or death decisions, and multi-million pharma investments. The fitness market requires a far more constrained model, and often does not need the avatar to look like the user, only to be influenced by the user. The commercial requirement (for example, in military, disaster and emergency service trainers as against dedicated medical treatment simulators) is often somewhere between the two. However, all can help to inform the development of appropriate physiological models for virtual humans across a wide variety of different application areas.

SENSES

The senses are how physical humans receive information from the physical world. As with physical humans, virtual humans rely on their senses to understand their world. Whilst some of those senses may be similar to, or the same as, human senses (such as seeing and hearing), others might be quite unique, for example, data. Within the context of a virtual human, the focus is primarily on how the digital entity interacts with others (human or AI) within a digital world, for example, a virtual world or training simulation, or through some other form of interface to the physical world. Given this, the main human senses which are relevant are:

- Sight, especially to 'see' the environment, or a webcam image of a user, or to be able to 'view' a map, diagram or video; and
- Hearing, especially for speech-to-text interaction with a user.

However, in addition to these senses, the virtual human very much has a sense of 'data', being able to receive information already in a digital format which it can then interpret either as pure information, or as an analogue for one of the senses, for example, the location/description of an object as an analogue for sight. Indeed, for a virtual human, all senses are experienced as data, but this may only fundamentally be in the same way that, for the human brain, all that is really being experienced are electrical nerve signals and biochemical messages – whatever 'sense' they may commonly be referred to as. People with conditions such as synesthesia and phantom limbs clearly show how even the brain can become confused and either treat one input as the 'wrong' sense or invent inputs for things that don't physically exist.

Sight

Computer vision (as the application of sight to computers is usually referred) has advanced since 2000 from basic lab work on feature recognition to applications embedded in smartphones, autonomous vehicles, face recognition systems, and military and industrial robotics. Given the virtual human use case, a key objective is to enable the virtual human to be able to hold a webcam conversation with a human user, such as in a human-to-virtual human Skype call. In these situations, being able to make sense of the visual imagery would enable the virtual human to:

- Recognise the user if known to them, or make deductions about the user (age, gender, ethnicity, 'style') if not;
- Mirror the human's body language and identify turn-taking cues and expressions, including when they might be getting bored or distracted – all helping to make the exchange more 'human';
- Detect emotional cues, which would enable the virtual human to show empathy and/or modify its conversation in order to better reach some specific goal; and
- Recognise objects or images that the user shows to the virtual human, and recognise how they are being manipulated, and perhaps even guide the user through a set of steps.

As examples of the research being done in these areas, Balcan et al. (2005) reports on the use of machine learning techniques to identify users using low-quality webcams. Ekenel et al. (2007) similarly describes the use of face recognition algorithms for a 'portable face recognition system that aims at environment-free face recognition'. Agulla et al. (2008) presents an educational application of the technology to prove user identity when accessing eLearning systems. Android phones introduced face recognition-based locking in 2012 in Android V4.0.

In terms of emotion detection, Benoit et al. (2009) has been investigating the detection of stress, and Pfister et al. (2011) the detection of truthfulness. Tan et al. (2010) presents a method for detecting 'face liveliness', here presented to differentiate between a spoofed still image and a real user, but with wider implications. Poh reflects much ongoing research by showing how a webcam can be used to 'extract the blood volume pulse from the facial regions … [and how] … heart rate (HR), respiratory rate, and HR variability (HRV, an index for cardiac autonomic activity) were subsequently quantified' (Poh et al., 2011). De and Saha (2015) provides a comparative study on different approaches of real time human emotion recognition based on facial expression detection. Ramirez et al. (2014) looks specifically at the use of skin colour (for example, blush) to inform emotional state, and Lozano-Monasor et al. (2014) looks at the detection of emotion (using Ekman's six basic emotions) from automated facial expression classification.

Hearing

Within the context of a virtual human, operating either as an avatar in a virtual world or as a robot, the need for the ability to hear and listen can be broken down into 3 main areas:

- Understanding speech (which is considered in more detail later in the Chapter 5);
- Identifying music and related artistic sounds; and
- Recognising ambient sounds, and identifying specific elements, for example, bird songs or warning sounds.

Services such as Shazam (https://www.shazam.com) have existed since the early 2000s and are able to take short sections of live audio captured by mobile phones and identify the piece of music playing. Shazam achieves greater than 95% accuracy on a 10-second or longer clip (Wang, 2003). If this capability was combined with a virtual human's ability to access an extensive song and review database, for example, Discogs (https://www.discogs.com/) with information on over 151 million tracks, a virtual human would be readily able to note which pieces of music are playing at any time, comment on them, and remember them as part of its episodic memory. It would also be able to use human responses to the music to categorise them as happy, sad, etc, and then, through machine learning techniques, identify an appropriate emotional response to a new piece of music (Lin et al., 2011), a technique known as Music Emotion Recognition (MER) (Yang and Chen, 2011). A more significant question is perhaps whether the virtual human could 'understand' or 'be moved by' the music, or show some other innate emotional reaction, and some work has been done in this area (Widmer, 1994). There is also a long history of using AIs to try and compose music (De Mantaras and Arcos, 2002).

As well as speech and music, a virtual human could be given the ability to understand the broader soundscape in either the virtual or physical world. For instance, Dufaux et al. (2000) describes a system that can identify transient sounds, such as glass breaking, car smashes, human screams, gunshots, explosions and other alarm sounds. Stowell et al. (2016), amongst others, is working on the identification of birdsong, and Lu et al. (2009) reports on the ability to classify the type of environment that the user is in purely from the background ambient sounds.

Other Human Senses

Smell and taste are human senses that have seen little application within the digital sphere, although there is work on electronic noses, smell generators, electronic tongues and even taste generators that could all be applied to a virtual human. There is significant work on touch as it applies to a human interacting with the virtual world (haptics), and force-feedback devices would give virtual humans a touch interface to the physical world. Touch within a virtual world is a commonly implemented feature, typically as a collision at the body level, or specific 'mouse-click' action, although there is some work on finer level touch detection and feedback. As discussed in this chapter, physiological models can be used to give virtual humans a sense of their own bodies, and this could even include a proprioceptive sense (the position and movement of the body) and an interoceptive sense of what is (notionally) going on inside their own virtual bodies. There is also no reason why a virtual human should be limited to human senses, as it could quite readily be interfaced with systems that detect radio waves or other physical phenomena.

Data as a Sense

As mentioned at the beginning of this chapter, for a virtual human, senses are all data in the same way that, for a human, they could be said to be all electrical impulses and biochemical signals. A virtual human's acquisition of data can be usefully broken down into three main elements: that from the virtual world, that from the physical world, and that from the cyber-world.

Data from a Virtual World

In most virtual worlds, the user's computer is being passed data by the virtual world servers that define what they can sense in the world – usually, the terrain, objects and avatars that they can see and hear. This data is dispatched as topology, texture images and audio file streams. The virtual human does not need to convert this information into visual and audio elements in order to understand it, it can directly consume the data. In fact, it can often by-pass a lot of the processing as, for example, the server might say that there is an object called 'Chair' at a particular location. The virtual human, if lazy, does not need to analyse the topology information, called the 'mesh', to confirm that the object does indeed have four legs and a flat surface and a back. It can just take it on trust that it is a chair. Of course, the untrusting virtual human, or any virtual human faced with an object called something like 'object3726', would need to analyse the topology, and maybe even the texture data to work out what the object is, but it still doesn't need to render it visually. However, as objects become more complex, say a large building, it might be quite challenging to work out from the topology what the building represents, in which case the virtual human could then render it to an image, run through a machine learning classifier and say 'ah, it's a church'.

Data from a Physical World

Data from the physical world are important for virtual humans that are occupying robot bodies, and for virtual human's communicating from the virtual to the physical world. Here, the virtual human can leverage the research into robotic sensors, much of which has just been described above.

Data from the Cyber-World

Finally, the data being received from the cyber-world (for example the Internet) would include every website, database, application programming interface (API) and data feed that the virtual human has access to. Again, the virtual human does not need to render a web page or display a news feed as text, it can consume the information directly. The virtual human may well develop a sense akin to taste or smell that lets it know whether this is a 'good' data feed or not, based on parameters such as data-rates, error-rates, recency, originality, authorship and veracity.

CONCLUSION

The interest of the film and games industry has seen significant improvements in the ability to model, animate and render digital avatars of the human body over the last few decades. Whilst the film industry can afford the time and processing requirements of highly accurate approaches, the games industry, as with the virtual human, needs to have methods which will work in real time, and, for the short term, is willing to accept the loss in fidelity that this often results in. In almost all aspects, designers have to be aware of the 'uncanny valley', and research has showed that users and viewers invariably prefer an avatar to be unrealistic rather than attempt to perfectly reproduce some aspect of the human body (or behaviour) but just fail. However, it does seem that within the next decade or so, real-time simulation of the appearance and animation of the human body will be sufficient to pass the 'Gold' Loebner Test, such that a user, over a system such as Skype, cannot tell whether they are watching a physical human, or a virtual one.

The chapter has also illustrated how all of the common human senses can be implemented for use by a virtual human, where the developments in robotics are often a key driver. Having considered how the appearance and senses of a human can be modelled in virtual form, the next chapter will examine how the behaviour and the associated cognitive processes of a human can also be modelled through computer software.

REFERENCES

Adabala, N. (2004). Rendering of skin and clothes. In Magnenat-Thalmann, N. & Thalmann D. (Eds.), *Handbook of Virtual Humans.* Chichester, UK: John Wiley & Sons. 353–372.)

Agulla, E. G., Rifón, L. A., Castro, J. L. A., & Mateo, C. G. (2008, July). Is my student at the other side? Applying biometric web authentication to e-learning environments. In *Advanced Learning Technologies, 2008. ICALT'08. Eighth IEEE International Conference on* (pp. 551–553). New York: IEEE.

Aliaga, C., O'Sullivan, C., Gutierrez, D., & Tamstorf, R. (2015). Sackcloth or silk?: the impact of appearance vs dynamics on the perception of animated cloth. In *Proceedings of the ACM SIGGRAPH Symposium on Applied Perception* (pp. 41–46). New York: ACM.

Balas, B., & Pacella, J. (2015). Artificial faces are harder to remember. *Computers in Human Behavior, 52*, 331–337.

Balcan, M. F., Blum, A., Choi, P. P., Lafferty, J., Pantano, B., Rwebangira, M. R., & Zhu, X. (2005, August). Person identification in webcam images: An application of semi-supervised learning. In *ICML 2005 Workshop on Learning with Partially Classified Training Data* (Vol. 2, p. 6).

Bartley, J., Forsyth, J., Pendse, P., Xin, D., Brown, G., Hagseth, P., & Hammond, T. (2013). World of workout: A contextual mobile RPG to encourage long term fitness. In *Proceedings of the Second ACM SIGSPATIAL International Workshop on the Use of GIS in Public Health* (pp. 60–67). New York: ACM. Available online http://healthgis.tamu.edu/ACMHealthGIS/2013/healthgis-09.pdf (Last accessed December 22, 2015).

Benoit, A., Bonnaud, L., Caplier, A., Ngo, P., Lawson, L., Trevisan, D. G., & Chanel, G. (2009). Multimodal focus attention and stress detection and feedback in an augmented driver simulator. *Personal and Ubiquitous Computing*, 13(1), 33–41.

Bertails, F., Audoly, B., Cani, M. P., Querleux, B., Leroy, F., & Lévêque, J. L. (2006). Super-helices for predicting the dynamics of natural hair. In *ACM Transactions on Graphics, 25*(3), 1180–1187.

Boulic, R., Thalmann, N. M., & Thalmann, D. (1990). A global human walking model with real-time kinematic personification. *The Visual Computer, 6*(6), 344–358.

Brooke, E. (2104). "8 Startups Trying To Help You Find Clothing That Fits" on the Fashionista blog (July 22, 2014). Available online http://fashionista.com/2014/07/8-tech-startups-tackling-clothing-fit.

Cao, Y., Tien, W. C., Faloutsos, P., & Pighin, F. (2005). Expressive speech-driven facial animation. *ACM Transactions on Graphics(TOG), 24*(4), 1283–1302.

Cassell, J., Vilhjálmsson, H. H., & Bickmore, T. (2004). BEAT: The behavior expression animation toolkit. In Cassell, J., Högni, H., Bickmore, T. (Eds.), *Life-Like Characters*. (pp. 163–185). Berlin, Germany: Springer.

Cirio, G., Lopez-Moreno, J., Miraut, D., & Otaduy, M. A. (2014). Yarn-level simulation of woven cloth. *ACM Transactions on Graphics (TOG), 33*(6), 207.

Daldegan, A., Thalmann, N. M., Kurihara, T., & Thalmann, D. (1993). An integrated system for modeling, animating and rendering hair. *Computer Graphics Forum, 12*(3), 211–221.

Daubert, K., Lensch, H. P. A., Heidrich, W., & Seidel, H. P. (2001) Efficient cloth modeling and rendering. In S. J. Gortler & K. Myszkowski. (Eds.), *Rendering Techniques*. (pp. 63–70). Eurographics. Vienna, Austria: Springer.

De, A., & Saha, A. (2015, March). A comparative study on different approaches of real time human emotion recognition based on facial expression detection. In *Computer Engineering and Applications (ICACEA), 2015 International Conference on Advances in* (pp. 483–487). New York: IEEE.

De Mantaras, R. L., & Arcos, J. L. (2002). AI and music: From composition to expressive performance. *AI magazine, 23*(3), 43.

Díaz-Zuccarini, V., Thiel, R., & Stroetmann, V. (2014). The European virtual physiological human initiative. In M. Rosenmöller., D. Whitehouse & P. Wilson. (Eds.), *Managing EHealth: From Vision to Reality* (pp. 244–258). London, UK: Palgrave Macmillan.

Dufaux, A., Besacier, L., Ansorge, M., & Pellandini, F. (2000, September). Automatic sound detection and recognition for noisy environment. In Signal Processing Conference, 2000 10th European (pp. 1–4). New York: IEEE.

Ekenel, H. K., Stallkamp, J., Gao, H., Fischer, M., & Stiefelhagen, R. (2007, July). Face recognition for smart interactions. In *Multimedia and Expo, 2007 IEEE International Conference on* (pp. 1007–1010). New York: IEEE.

Ezzat, T., Geiger, G., & Poggio, T. (2002). Trainable videorealistic speech animation. *ACM Transactions on Graphics (TOG), 21*(3), 388–398.

Fan, S., Ng, T.-T., Herberg, J., Koenig, B. L., & Xin, S. (2012). Real or fake?: Human judgments about photographs and computer-generated images of faces. *Proceedings of SIGGRAPH Asia 2012 Technical Briefs, 17*(18).

Farid, H., & Bravo, M. J. (2012). Perceptual discrimination of computer generated and photographic faces. *Digital Investigation, 8*(3), 226–235.

Frâncu, M., & Moldoveanu, F. (2015). Virtual try on systems for clothes: Issues and solutions. *UPB Scientific Bulletin, Series C, 77*(4), 31–44.

Hadap, S., & Magnenat-Thalmann, N. (2001). Modeling dynamic hair as a continuum. In *Computer Graphics Forum, 20*(3), 329–338.

Kofránek, J., Mateják, M., Privitzer, P., Tribula, M., Kulhánek, T., Silar, J., & Pecinovský, R. (2013). HumMod-golem edition: Large scale model of integrative physiology for virtual patient simulators. *Proceedings of World Congress in Computer Science 2013 (WORLDCOMP'13), International Conference on Modeling, Simulation and Visualisation Methods (MSV'13)* 182–188.

Kramar, V., Korhonen, M., & Sergeev, Y. (2013). Particularities of visualisation of medical and wellness data through a digital patient avatar. In *Open Innovations Association (FRUCT), 2013 14th Conference of* (pp. 45–56). IEEE.

Kshirsagar, S., Egges, A., & Garchery, S. (2004). Expressive speech animation and facial communication. In Magnenat-Thalmann, N. & Thalmann D. (Eds.), *Handbook of virtual Humans* (pp 230–259). Chichester, UK: John Wiley & Sons.

LeBlanc, A. M., Turner, R., & Thalmann, D. (1991). Rendering hair using pixel blending and shadow buffers. *Computer Animation and Virtual Worlds, 2*(3), 92–97.

Lin, Y. C., Yang, Y. H., & Chen, H. H. (2011). Exploiting online music tags for music emotion classification. *ACM Transactions on Multimedia Computing, Communications, and Applications* (*TOMM*), *7*(1), 26.

Loper, M., Mahmood, N., Romero, J., Pons-Moll, G., & Black, M. J. (2015). SMPL: A skinned multi-person linear model. *ACM Transactions on Graphics (TOG), 34*(6), 248.

Lopez, C. (2014). Latest "Virtual Battle Space" release adds realism to scenarios, avatars on www.army.mil Available online http://www.army.mil/article/123316/Latest__Virtual_Battle_Space__release_adds_realism_to_scenarios__avatars/?from=RSS.

Lozano-Monasor, E., López, M. T., Fernández-Caballero, A., & Vigo-Bustos, F. (2014). Facial expression recognition from webcam based on active shape models and support vector machines. In *Ambient Assisted Living and Daily Activities* (pp. 147–154). London, UK: Springer International Publishing.

Lu, H., Pan, W., Lane, N. D., Choudhury, T., & Campbell, A. T. (2009, June). SoundSense: Scalable sound sensing for people-centric applications on mobile phones. In *Proceedings of the 7th International Conference on Mobile Systems, Applications, and Services* (pp. 165–178). New York: ACM.

MacDorman, K. F., Green, R. D., Ho, C. C., & Koch, C. T. (2009). Too real for comfort? Uncanny responses to computer generated faces. *Computers in Human Behavior, 25*(3), 695–710.

Miguel, E., Feng, A., Xu, Y., Shapiro, A., Tamstorf, R., Bradley, D., Schvartzman, S. C., & Marschner, S. (2014). Towards cloth-manipulating characters. *Computer Animation and Social Agents, 3*, 1.

Mori, M. (1970). The uncanny valley. *Energy, 7*(4), 33–35.

Nagano, K., Fyffe, G., Alexander, O., Barbiç, J., Li, H., Ghosh, A., & Debevec, P. E. (2015). Skin microstructure deformation with displacement map convolution. *ACM Transactions on Graphics, 34*(4), 109–111.

Patrick, D., Bangay, S., & Lobb, A. (2004). Modelling and rendering techniques for African hairstyles. In *Proceedings of the 3rd International Conference on Computer Graphics, Virtual Reality, Visualisation and Interaction in Africa*, New York: ACM. 115–124.

Pfister, T., Li, X., Zhao, G., & Pietikäinen, M. (2011, November). Recognising spontaneous facial micro-expressions. In Computer Vision (ICCV), 2011 *IEEE International Conference on* (pp. 1449–1456). New York: IEEE.

Poh, M. Z., McDuff, D. J., & Picard, R. W. (2011). Advancements in noncontact, multiparameter physiological measurements using a webcam. *IEEE Transactions on Biomedical Engineering, 58*(1), 7–11.

Pons-Moll, G., Romero, J., Mahmood, N., & Black, M. J. (2015). Dyna: A model of dynamic human shape in motion. *ACM Transactions on Graphics (TOG), 34*(4), 120.

Ramirez, G., Fuentes, O., Crites, S. L., Jimenez, M., & Ordonez, J. (2014, June). Color analysis of facial skin: Detection of emotional state. In *Computer Vision and Pattern Recognition Workshops* (*CVPRW*), 2014 IEEE Conference on (pp. 474–479). New York: IEEE.

Rosenblum, R. E., Carlson, W. E., & Tripp, E. (1991). Simulating the structure and dynamics of human hair: Modelling, rendering and animation. *Computer Animation and Virtual Worlds, 2*(4), 141–148.

Sattler, M., Sarlette, R., & Klein, R. (2003). Efficient and realistic visualization of cloth. In S. J. Gortler & K. Myszkowski. (Eds.), *Rendering Techniques* (pp. 167–178). Eurographics. Vienna, Austria: Springer.

Seo, H. (2004). *Anthropometric Body Modelling Handbook of Virtual Humans.* (pp. 75–98). Chichester, UK: John Wiley & Sons.

Sintorn, E., & Assarsson, U. (2008). Real-time approximate sorting for self shadowing and transparency in hair rendering. In *Proceedings of the 2008 symposium on Interactive 3D graphics and games* (pp. 157–162) New York: ACM.

Staples, A. (2012). Subsurface scattering: Skin shaders for Poser 9/Poser Pro. Penultimate Harn website. Available online http://www.penultimateharn.com/3dgallery/SubsurfaceScattering01.pdf.

Stowell, D., Wood, M., Stylianou, Y., & Glotin, H. (2016). Bird detection in audio: A survey and a challenge. *arXiv preprint arXiv*:1608.03417.

Stuart, K. (2015). "Photorealism—The future of video game visuals," Guardian website (February 12, 2015). Available online http://www.theguardian.com/technology/2015/feb/12/future-of-video-gaming-visuals-nvidia-rendering

Tan, X., Li, Y., Liu, J., & Jiang, L. (2010). Face liveness detection from a single image with sparse low rank bilinear discriminative model. In *Computer Vision–ECCV 2010* (pp. 504–517). Berlin, Germany: Springer.

Tolani, D., Goswami, A., & Badler, N. I. (2000). Real-time inverse kinematics techniques for anthropomorphic limbs. *Graphical Models, 62*(5), 353–388.

UCL. (2015). The virtual physiological human. Available online https://www.ucl.ac.uk/news/vph.doc.

Visage. (2012). MPEG-4 Face and body animation (MPEG-4 FBA) an overview. Available online http://www.visagetechnologies.com/uploads/2012/08/MPEG-4FBAOverview.pdf.

Volino, P., & Cordier, F. (2004). "Cloth simulation" in *Handbook of Virtual Humans* (pp. 192–229). Chichester, UK: John Wiley & Sons.

Widmer, G. (1994, October). The synergy of music theory and AI: Learning multi-level expressive interpretation. In *Proceedings of the National Conference on Artificial Intelligence* (pp. 114–114). Chichester, UK: John Wiley & Sons Ltd.

Wang, A. (2003, October). An industrial strength audio search algorithm. In *Ismir* (Vol. 2003, pp. 7–13). Canada.

Ward, K., Bertails, F., Kim, T. Y., Marschner, S. R., Cani, M. P., & Lin, M. C. (2007). A survey on hair modeling: Styling, simulation, and rendering. *IEEE Transactions on Visualization and Computer Graphics, 13*(2), 213–234.

Wu, K., & Yuksel, C. (2017). Real-time fiber-level cloth rendering. In *Proceedings of the 21st ACM SIGGRAPH Symposium on Interactive 3D Graphics and Games*. New York: ACM.

Yang, Y. H., & Chen, H. H. (2011). Ranking-based emotion recognition for music organization and retrieval. *IEEE Transactions on Audio, Speech, and Language Processing, 19*(4), 762–774.

Yang, R., Li, D., Hu, Y., Liu, J., & Hu, D. (2015). Rapid skin deformation algorithm of the three-dimensional human body. *7th International Conference on Intelligent Human-Machine Systems and Cybernetics* (pp. 244–247) Hangzhou, China.

Zhang, M., Lin, L., Pan, Z., & Xiang, N. (2015). Topology-independent 3D garment fitting for virtual clothing. *Multimedia Tools and Applications, 74*(9), 3137–3153.

CHAPTER 4

Mind

THIS CHAPTER COMPLEMENTS THE study of the 'body' and senses technologies in Chapter 3 by examining the different technologies and approaches involved in the creation of the mind or brain of a virtual human. The chapter will consider current research in each of these areas and anticipate the future directions that may develop over the coming decade. Issues connected to communications, including natural language, will be examined in Chapter 5. The section headings follow the elements shown in the Virtual Human Component Architecture at Figure II.1.

UNDERSTANDING WHAT CONSTITUTES THE MIND

In considering how to build a virtual human, the mind represents that part of the virtual human which is receiving sensory information from a physical, virtual or cyber world, deciding what to do in response to both that information and its own drives and needs, and then sending the instructions to the virtual body to perform some action. In undertaking this sense-decide-action loop, the virtual human will begin to create memories, build up beliefs, learn information and processes, and even create its own goals. In order to be accepted as a virtual 'human' rather than just a virtual robot it is also to be expected that the virtual human will display (and ideally feel) emotion and empathy, have moods, and that a personality will emerge. There are, hence, two closely interlinked processes involved: the sense-decide-action loop, which links the virtual human directly to the external world, and the virtual human's internal inner narrative and development. These are shown in Figure 4.1.

In examining the detail of the processes involved, it is important in this chapter to consider how the mind receives the information that is coming

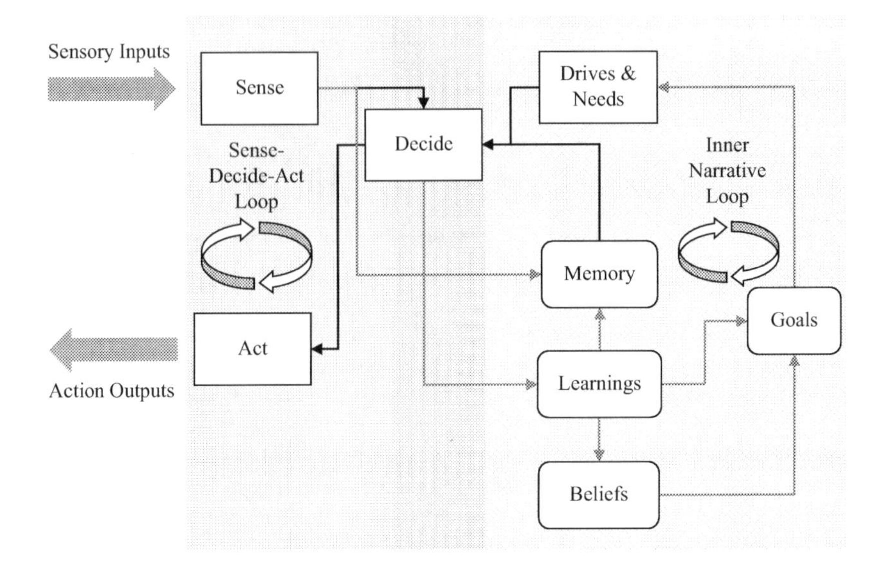

FIGURE 4.1 The key mind processes of a virtual human.

from its senses, the processes of perception and attention, followed by the activity of appraisal (how the mind evaluates those sensory inputs), and then how that appraisal influences the mind's emotional state and overall mood, and, ultimately, influences its personality. Having established a reactive chain, the more proactive elements of the virtual human's mind can be considered. This involves the way in which the mind is motivated and subsequently establishes goals based on its needs, and how it derives plans in order to fulfil those goals. In executing a plan, the virtual human will need to be able to reason, to make decisions and to solve problems. Achieving any of these will require information, which, in turn, demands that the memory is effective, and if the virtual human is to warrant the label of 'intelligent', then it would also be expected to learn from its actions, recording new memories and procedures. At the highest level, it would be expected that these experiences would lead to changes in the virtual human at a meta-level, changing its personality, motivations, needs and goals. These key components of this active mind and associated mental processes will now be considered.

PERCEPTION

Perception is the process by which an entity receives a signal from some sensory system, for example, sight or hearing, and generates an internal representation of what is being perceived, and then consciously experiences it (Prinz, 2006). However, what the brain perceives may be different to what is really there – what some have defined as an 'internal external-reality' (Lilly, 2010). Such perception effects may range from simple visual illusions to making mistaken assumptions about people. This process of perception is initially a subconscious one, until followed by a more conscious attention process where the brain focuses on the most interesting pieces of information. Cohen and Feigenbaum (2014) describe how in vision algorithms there are 'two aspects of human visual perception. First, the eye itself has multiple levels of resolution … second, humans naturally attend to areas of high information and ignore less interesting areas. This is called selective attention'.

Whilst the first is primarily a computer vision task, it is the second aspect, selective attention, which needs to be implemented within the virtual human mind. Crowder and Friess (2012) identifies that perception can also be influenced by our emotional state, so it is not just a mechanical case of what is seen and seems to hold information, but also the (often subconscious) mind putting its own interpretation on what is seen – and this may be influenced by past experience and culture, as well as by a person's current mood.

Implementing perception is often a by-product of the connecting of a sensor to the virtual human, or other target system. However, it is one area where creating a more 'sophisticated' virtual human is possibly about degrading the initially 'perfect' interface to create something which mirrors the flaws of human perception.

ATTENTION

Attention in AI, and particularly robotics, is better defined than perception. Where a sensor has a limited field of view, then attention is the directing of those sensors onto the area of interest – and is found in the gaze control of robotic heads as they follow the conversation between two people speaking, or the movement of an object of interest. As well as the heightened interest in and analysis of the subject of the attention, there is also often a corresponding reduction in the awareness of activity outside of the area (topical or spatial) of attention.

Similar effects are seen purely in terms of how a human applies brain-power to making sense of a complex scene. One of the best known examples of such topical attention, referred to as 'inattention blindness', is the video experiment by Simons and Chabris (1999), which clearly shows how people focusing on one activity may completely miss another activity in the same visual scene.

Kismet (Breazeal, 2004) is one of the best known early robots that was designed around human interaction (see Figure 4.2). It has exaggerated facial features to better signal emotion, and head and gaze tracking to not only focus on a person talking or object of interest but to make head gestures and facial expressions to optimize their distance from the robot. A more recent example is Muecas, described by Cid (2014).

Agüero et al. (2012) describes how robotic visual attention mechanisms are divided into covert (where the attention is paid on a part of a usually fixed or uncontrolled visual stream), and overt (where the camera can be actively moved to focus on the item(s) of interest). Agüero also describes three different overt attention mechanisms which were tested within the context of the RoboCup: one based on round-robin sharing, another

FIGURE 4.2 Kismet robot. (Source: Wikicommons. © Jared C. Benedict.)

based on dynamic salience and one with fixed pattern camera movements. However, Borji et al. (2012, 444) highlights the dangers of attention algorithms, and there is a notable overlap with perception:

> The main concern in modelling [attention] saliency is how, when, and based on what, to select salient image regions. It is often assumed that attention is attracted by salient stimuli or events in the visual array. While this is the case, it is also known that a larger portion of attentional behavior comes from ongoing task inferences which dynamically change and are dependent on the algorithm of the task.

In other words, attention is as much a cognitive task as it is an image processing task.

In working with virtual humans in virtual worlds, the authors have found that even the most basic attention routines can dramatically improve the humanness of an avatar. If a user approaches an avatar (which they assume is human, as they have no information yet to the contrary) and speak to it, and it replies without turning to face them, then they may immediately suspect that it is a bot – or just very rude! However, if the avatar naturally turns to them before replying—as most humans would do with their avatar when first meeting someone—then the illusion of humanness is maintained.

APPRAISAL, EMOTION AND MOOD

Within the context of AI research, appraisal theory seeks to identify how 'a person's emotional experience is determined by his/her subjective assessment and interpretation of internal or external events' (Si, 2015). As such, the same event trigger can be appraised in different ways at different times, and so have a different emotional effect. In cognition terms, emotion is usually defined in terms of the instant (millisecond to second duration) response to an event. That emotion may then influence (and be influenced by) the longer term 'mood' of the person, which will, in turn, be affected by their overall personality (Wilson, 2000).

Emotions can be 'expressed' by a virtual human's avatar in a number of ways including: gesture, facial expression, sounds, speech, movement, and physiological changes (for example, sweat, blush, etc.) (Izard, 2013).

Models of emotion have been categorised as Adaptional, Discrete, Motivational, Dimensional, Appraisal and Constructivist (Scherer et al., 2010).

Of these, the Discrete and Appraisal categories have produced the two best known models of emotion. Ekman (1989) is representative of the Discrete emotion category and produced a commonly accepted set of 'primary' emotions:

- Happiness,
- Sadness,
- Anger,
- Fear,
- Disgust, and
- Surprise (although surprise is sometimes treated as a separate category).

Other secondary emotions can be considered as combinations of these.

The Ortony, Clore and Collins (OCC) model (Ortony, 1990) is of the appraisal type and has been influential in AI research, and considers emotions as valenced reactions to events, people and objects. Derived from OCC is the E-AI Emotional Architecture model developed by Wolverhampton University (Slater and Burden, 2009) and which has been implemented in part by Daden in the Halo virtual human discussed in Chapter 8. It includes the following stages, which are reflected in many other models:

- Detection of stimulus (Sensation/perception),
- Appraisal of stimulus – against an Emotion Alert Database (EAD),
- Unconscious reaction,
- Physiological changes,
- Motivation to act, and
- Conscious realization (Feelings).

The WASABI Affect Simulation Architecture for Believable Interactivity (WASABI) system (Becker-Asano, 2014), discussed in more detail in Chapter 6, uses a 3D pleasure-arousal-dominance (PAD) space to represent secondary emotions – bringing together both appraisal and dimensional approaches. As with many affective systems, the dominant emotion has a decay rate that returns the virtual human to a neutral state in the absence of any further stimuli.

As an example of such systems in action, Salichs and Malfaz (2012) have created a social robot called Maggie which implements an appraisal-based emotional model and closely ties it into a motivation, drives and needs model. Adam and Lorini (2014), also describe the emotional and empathic design of a social companion robot based on the ProLog Emotional Intelligent Agent Designer (PLEIAD) system, and link emotion to both motivation and Belief-Desire-Intention (BDI) models (see below). Adam's emphasis on empathic design highlights that a virtual human or robot ideally needs to be able to not only show emotion but also to detect emotion in others and adjust its behaviour accordingly.

Again, the authors have found that implementing a relatively basic emotional architecture into an avatar in a virtual world (as described by Slater and Burden, 2009) creates a significantly more human-like avatar. A challenge, though, is to make the emotional behaviour subtle. It is all too easy to create significant displays of happiness, sadness, fear, disgust and so on for the most minor of triggers, but far harder to communicate more subdued responses. Luckily, in many virtual environments, over-the-top behaviour is the norm and so the virtual human can fit right in.

PERSONALITY

Bogdanovych et al. (2015) consider personality to be one of the key features of believable agents, stating that 'personality infuses everything a character does, from the way they talk and move to the way they think. What makes characters interesting are their unique ways of doing things. Personality is about the unique and not the general'.

By embedding a personality model within a virtual human, the finer levels of behaviour can be inherited from the personality, rather than having to be defined explicitly for each virtual human. Just as with emotional models, there are several different categories of personality models, including trait/factor-based and role-based.

Puică and Florea (2013) uses only two dimensions (Eysenck's neuroticism and extraversion) but uses this to derive four temperament types: Sanguine, Choleric, Melancholic and Phlegmatic). McRorie et al. (2012) uses a three-factor model (psychoticism, neuroticism and extraversion) based on Eysenck. Wilson (2000) also uses a three-parameter personality model, based on three axes of Extroversion, Fear and Aggression (EFA space). Wilson also relates personality to mood, emotion and motivation.

Doce et al. (2010) uses the Five Factor Model (openness to experience, conscientiousness, extraversion, agreeableness, and neuroticism) and states that 'using an explicit personality model, based on well-established trait theories, such as the Five Factor Model (FFM) of personality, provides a better tool to easily create coherent and different [artificial] personalities'.

This Five Factor Model is also used by Pérez-Pinillos et al. (2013) and Bogdanovych et al. (2015) in developing their virtual agents.

One of the most well-known personality models is that of Belbin (2012), a role-based model which identifies nine personality roles within teams: Plant, Resource Investigator, Coordinator, Shaper, Monitor Evaluator, Team-worker, Implementer, Completer/Finisher, Specialist. Aguilar (2007) describes a system to use Intelligent Virtual Agents (IVAs) as part of Team Training exercises, particularly in the 'substitution of missing team members (Teammate Agents) to promote the practice of team tasks in relation to functional roles'. To achieve this, the agents had personalities modelled using Belbin's team roles and implemented via the Behaviour Expression Animation Toolkit (BEAT) system. However, Sansonnet and Bouchet (2013, 1) identifies potential limitations of current approaches stating that:

> Most research works on the computational implementation of psychological phenomena usually fail to take into account two key notions: coverage, as they often focus only on a small subset of psychological phenomena (for example considering few traits), and comprehensiveness, because they resort to procedural implementations (such as hard-coded rules) therefore excluding experts (i.e., psychologists) from the agent's behavior design process.

The well known trait-based 16PF model (Cattell et al., 1970) offers the most dimensions of those models considered here, measuring the strengths of 16 traits, expressed as spectrum extremes, ranging from shy/bold to expedient/rule-conscious, although the increased resolution does not necessarily make it a better or more accurate model.

Having effective personality models not only enhances believability (Bogdanovych et al., 2015), but also reduces the effort needed to create a large number of virtual humans within simulations, such as disaster simulations (Kefalas et al., 2014). As Lim et al. (2012) notes, existing techniques based on 'finite states and rules can be very tedious' and that 'it would be desirable for some of this burden to be offloaded from the designer's shoulder' by being able to define (or randomly generate) a range of personalities, and then have the NPCs behave accordingly.

A significant challenge in implementing personality is having suitable ways in which to display those personality traits. For instance, one of the Digital Immortality systems described in Chapter 12 collects a large amount of personality data using industry standard personality surveys, but since the 'digital immortal' is a very crude 2D head and shoulders avatar run by a fairly basic natural language system, the virtual persona has minimal ways of displaying its personality traits. One of the authors' current areas of study is how different personality traits may be manifest through different affordances of the virtual human.

MOTIVATION, GOALS AND PLANNING

In order to model a virtual human, an agent must have a set of motivations and goals to give direction and meaning to its actions, and to plan sequences of action. The latter element of this is the subject of much study, particularly within the context of autonomous robots, but tends to be limited to short term goals. The virtual human needs a longer-term set of goals, which could also be called motivations, but which could potentially be managed using a goal-based framework.

Salichs and Malfaz (2012) describe a general motivation model based on Lorenz's hydraulic model of motivation (Lorenz and Leyhausen, 1973). In this model, motivation depends on the balance between a stimulus and a need/drive. If a need/drive is low (for example, not feeling hungry) then the stimulus would have to be very great to have an effect (such as eating a wonderful pastry), whereas if the drive is high (for example feeling very hungry), then the stimulus can be a very low (for example a McDonalds).

In AI research, one of the most common motivation/planning models is the Belief-Desire-Intentions (BDI) model (Rao and Georgeff 1991). The AI has a set of assumptions (beliefs) about the current state of the world (and itself), a set of goals (called desires), and then a set of possible options or plans (which become intentions once chosen) about how to achieve those goals. The system then dynamically assesses the gaps between beliefs and desires, chooses the best option or plan as its intention, and then enacts that plan until either the desire is realized, or another desire has to be handled, or the plan hits a problem – when it then starts to re-plan. Agentspeak(L) is a formal language to implement a BDI system (Rahwan et al., 2004).

One possible approach to the issue of longer-term goals is to use a model of human motivation, such as Maslow's hierarchy of needs (Maslow, 1954), which, whilst contested, has a simplicity which can help when considering how to code such behaviours. Here, needs (the fulfilment of which are motivations) are not only categorised but also ranked, so that the lower levels needs must be satisfied before a person (or AI) will spend time on higher level needs. The standard Maslow's hierarchy of needs (in decreasing order of priority) is:

7. **Self-Actualization needs** – realizing personal potential, self-fulfilment, seeking personal growth and peak experiences.
6. **Aesthetic needs** – appreciation and search for beauty, balance, form, etc.
5. **Cognitive needs** – knowledge, meaning, etc.
4. **Esteem needs** – self-esteem, achievement, mastery, independence, status, dominance, prestige, managerial responsibility, etc.
3. **Love and belongingness needs** – friendship, intimacy, affection and love, – from work group, family, friends, romantic relationships.
2. **Safety needs** – protection from elements, security, order, law, stability, etc.
1. **Biological and Physiological needs** – air, food, drink, shelter, warmth, sex, sleep, etc.

Andriamasinoro (2004) describe how Maslow's hierarchy can be used to provide motivation for a hybrid artificial agent, and a similar model is implemented in the FAtiMA-PSI architecture described in Chapter 6.

Another approach (Salgado et al., 2012) is the Multi-level Darwinist Brain. Here, the AI maintains three models: one of the external world state, one of its own current state, and a third of its own satisfaction (which again could be based on a Maslow-type model). The approach then uses a genetic algorithm approach to try multiple options from a library (and variations on them) to see which best results in an optimised world, internal and satisfaction model, and that option is then implemented.

The higher the level of goals and motivations (and the means to achieve them) that can be given to a virtual human, then the less detailed 'programming' is required for it to carry out a set of tasks. In a very different use case, a virtual tutor with motivations around helping its students learn and grow and keeping them emotionally stable and motivated would be better able to personalise the delivery of a course of learning and be better able to support a student over the long term without much additional lower level coding. One of the challenges, though, of developing more complex or sophisticated motivation models with a wide scope is the risk of emergent and unexpected/unwanted behaviour as a large number of rules, conditions and options are evaluated.

In the authors' work with Halo (see Chapter 8), implementing even a basic motivational model meant that her behaviour became, at a practical level, far less mechanistic. Rather than waiting for her next command or the next visitor, she could be found walking in her garden, sat in her lounge, or even 'fast asleep' in her bed. The illusion of her living a life had started.

DECISION-MAKING, PROBLEM-SOLVING AND REASONING

Physical humans are faced with making decisions, solving problems and reasoning about things on a whole variety of scales. For example, from second-by-second motor-control decisions to moving hands to lift things up or placing fingertips on computer keyboards, through deciding what to do for the day and or what to have for lunch, and then working out how to open a stuck lid on a jar, to trying to evaluate political or ethical arguments, solving problems at work or deciding about a change in job. A virtual human, when faced with a problem or decision of whatever nature, needs the same ability to reason, problem-solve or to choose a course of action, preferably the 'best' course of action. What represents (to the virtual human) the 'best' course or the right decision will, of course, depend on its motivations and goals, as described above. Indeed, systems already mentioned (for example, BDI) already have decision-making components within them.

The decision-making process built into the Belief-Desire-Intentions (BDI) model (Rao and Georgeff 1991) is based on an evaluation of which plan meets the needs of the current goals and world-state. However, the model is heavily codified and whilst well suited to simple agent tasks may be problematic to apply to more real-world 'grey' decision-making situations. Pomerol (1997) identifies a wider range of decision-making techniques that an AI could make use of including: utility functions, pattern matching, expert systems, numerical methods, neural networks, and case-based reasoning. Another approach has been to try and give systems a level of common sense to help in reasoning and decision-making. Pease et al. (2000) describes the US Defense Advanced Research Projects Agency (DARPA)'s High Performance Knowledge Bases (HPKB) project, which focused on creating very large and fast knowledge bases for common sense reasoning in the domains of warfare and geography. Davis and Morgenstern (2004, 2) describes the challenges in trying to create such systems, as to do so one must:

> (1) develop a formal language that is sufficiently powerful and expressive; (2) capture the many millions of facts that people know and reason with; (3) correctly encode this information as sentences in a logic; and (4) construct a system that will use its knowledge efficiently.

The Open Mind Common Sense Project described by Singh (2012) is another example of a common sense-based system. This investigation into general-purpose reasoning and decision-making systems, particularly when based on common sense, leads naturally into the research on Artificial General Intelligence which is discussed in more detail in Chapter 13.

In many cases, a virtual human will be operating alongside a physical human, and the decision-making process may be shared. Parasuraman et al., (2000) provide a useful model for different levels of human and autonomous decision-making, and a key issue in current research (and practice) is in where to create a balance between human and machine decision-making and automation. There is also the danger that having the computer announce the solution/decision can make it seem more certain and 'right' than hearing it from a human, and so it may be less subject to challenge.

World Models

An inherent requirement for a virtual human to be able to decide, problem solve, reason and act on the world in a coherent way is that it should have a mental model of the world as it is. As Johnson-Laird (2005) describes it, '[m]ental models represent entities and persons, events and processes, and the operations of complex systems'. Jones et al. (2011) expands:

> People must know about their environment so they can exist within it (Moore and Golledge 1976). Mental models are conceived of as a cognitive structure that forms the basis of reasoning, decision making, and, with the limitations also observed in the attitudes, literature, behavior. They are constructed by individuals based on their personal life experiences, perceptions, and understandings of the world. They provide the mechanism through which new information is filtered and stored.

In order to deal with hypothetical situations and thought exercises (e.g., imagining Jane has two apples), the virtual human may need multiple world maps and models, one of which deals with the physical world (and perhaps others for different virtual worlds and other realms) and one or more for different hypothetical or imagined worlds. Such world models are likely to contain:

- Spatial relationships;
- Social relationships;
- Models of object interactions (e.g., the laws of physics);
- Models of social interactions; and
- Models of psychological behaviours.

Whilst the semantic memory will contain much of the background information required, the world model will also need its own spatial and social graph, and the models to support the interactions and behaviours within it. The spatial model is relatively straightforward, but the psychological and social model will reflect in many ways the concept of 'theory of mind' (Goldman, 2012).

Theory of mind posits that humans maintain a mental model of how other people behave and what they know. So, if subject S is told that person A sees object B being hidden in a drawer, but person A then leaves whilst person C moves object B to a shelf, then if asked, subject S would be expected to say that person A would still assume that the object was in the drawer. Various researchers (Perner, 1991) have found that such a theory of mind only develops in children at around the age of 4 years – before that, they only act on what they themselves know – so they would say that person A would say that object B is now on the shelf – as that is where they (the child) now know it to be. A theory of mind has even been observed in Chimpanzees – but not in any other lower monkeys (Frith and Frith, 2005).

It would be reasonable to expect that any virtual human would have a 'theory of mind' capability developed for it, maintaining (or creating as needed) a mental model of the world as seen by each participant in order to answer such questions and to reason and decide about the world and imagined situations. In fact, the virtual human would only be able to correctly parse and understand a scenario description such as that above if it had been programmed with at least a basic theory of mind. But the creation of such capability would be, like almost every other feature, a virtual human, a conscious act by its creators.

MEMORY

Memory is a central element of physical humans, and so it should be for virtual humans. As with many capabilities, physical humans only appreciate the role and importance of memory when it becomes impaired. The increasing prevalence of, and interest in, dementia highlights just some of the different ways in which memory can fail, and the quite devastating effects that can have on quality of life, and even humanness.

Rather than address memory in computer terms (RAM, ROM, hard disc), most AI research considers memory in human terms. Human memory is divided first into short-term (few seconds, current task) and long-term (everything else) memory (Vernon et al., 2015). The short-term (or working) memory is often referred to as a scratch pad, whiteboard, blackboard or workspace for whatever task or issue the virtual human is currently having to deal with. Baddeley (2012) provides a useful review of the evolution of working memory models. The dominant model appears to split working memory into three elements: a speech/sound loop (the phonological loop – for example, remembering a PIN or phone number) for about half-a-dozen items, a visual-spatial sketchpad (of a few seconds duration), and a central executive (whose roles include focusing attention, dividing attention, task switching, and managing the interface to long-term memory).
Long-term memory is usually divided into:

- Declarative memory (about things, people, concepts); and
- Procedural memory (how to do things).

Declarative memory is, in turn, split into:

- Episodic (things that have happened to the person or virtual human); and
- Semantic (knowledge about things, people, places, concepts and their relationships to each other).

Stachowicz and Kruijff (2012) describe an 'approach to providing a cognitive robot with a long-term memory of experiences – a memory, inspired by the concept of episodic memory (in humans) or episodic-like memory (in animals), respectively'. Memory not only provides the means to store experiences, but also allows them to be integrated into more abstract constructs (i.e., semantic memory), and to recall the content. Schank (1990) examines how storytelling and memory overlap and reinforce each other, and Kelley (2014) discusses the losing of memories and looks at how episodic memories can be 'pruned' in a method analogous to dreaming. Both of these approaches begin to move memory out of a very mechanistic domain and one closer to how a virtual human with an ever-changing set of memories might be expected to behave.

In terms of how information is encoded within AI memory, semantic networks and maps, particularly those derived from the Semantic Web (Berners-Lee, 2001) and using standards such as the Resource Descriptor Framework (RDF) (W3C, 2014), are emerging as a common approach as they seem to match the connectedness of human memories.

The development of virtual human memory seems to be developing more or less in step with our understanding of human memory. It is notable that there is even interest within the academic community in developing less than perfect memories to simulate the fallibility of human memory and to better understand conditions such as Alzheimer's (Kanov, 2017).

LEARNING

Whilst Chapter 11 will consider how virtual humans may be used in human education, it is also important that virtual humans can learn for themselves. As discussed in Chapter 1, learning is seen as a key element of intelligence, and a virtual human ultimately needs to be capable of the same (or similar) types of formal and informal learning as physical humans.

In the context of the development of virtual humans, learning can be seen as either 'passive' learning – where the virtual human acquires knowledge from other sources (for example, reading an RSS feed), or 'active' learning, where the virtual human tries things out to see if they work. This latter approach is often used in AI research in the form of reward functions within reinforcement learning (Sutton and Barto, 1998) – where the agent learns to carry out the actions, or combination of actions, that maximize particular rewards, although Mataric (1994) highlights that performance can be poor in situations with multiple goals, noisy states and inconsistent reinforcement.

Table 4.1 illustrates seven methods of learning described by Hopgood (2011), along with examples of how they could apply to a virtual human.

Having virtual humans that can learn could mean that there will be a significantly reduced need for programmers, and that instead a virtual human can be taught more by a subject matter expert, or even learn on its own. This creates a significant efficiency multiplier effect; a 'generic' virtual human with good learning capability could, quite rapidly, become a 'specialist' virtual human in one or more subjects or processes. As 'unskilled' virtual humans could be in almost limitless supply and could learn or be trained to undertake a wide variety of roles, from the trivial, such as NPCs

TABLE 4.1 Seven Methods of Virtual Human Learning

Method	Examples
Rote	Converting an RSS news feed, or text book page, into a set of RDF triples in the long term semantic memory, such as is done by the ConTEXT system (Khalili et al., 2014).
Advice	Using humans to do a task to provide a robot with a template to follow is a common approach in robotics (e.g., Rozo et al., 2013), often referred to as Learning from Demonstration (LfD), and can be applied in virtual environments (McCormick et al., 2014).
Induction	Induction learning is similar to that found in most machine learning techniques, where the computer is trained on a set of examples (e.g., images or texts) and can then use these examples to identify a new image or piece of text (Kiesel, 2005).
Analogy	Perhaps one of the more challenging approaches for a virtual human since it must understand how two very different examples can actually be closely related.
Explanation	The virtual human tries to develop the explanations for why things happen. The PRODIGY system (Minton et al., 1989) includes an implementation of an explanation-based learning system.
Case-based	The virtual human acquires sets of cases that describe particular situations and particular solutions, and then when faced with a new problem finds the best matching solution. Anthony et al. (2017) describes how a case-based reasoning system was used with an intelligent agent to provide support for software programmers.
Explorative	Explorative learning is probably best exemplified by genetic learning algorithms. Here, to find a solution, the virtual human tries a variety of parameters at random, and then taking the best performing solution 'breeds' a new set, which it tries again, selects and breeds again and so on until it finds the best solution (Bredeche et al., 2012). Of course, there is no reason why a virtual human's 'exploration' needs to be conducted in its 'normal' virtual world – it could readily be provided with a simulation of the simulation to try things out in first – so-called Projective Simulation (Mautner et al., 2014).

in video games, to the vital, such as caring for an aging population, their impact on the need for human resource could be immense.

IMAGINATION AND CREATIVITY

Defining terms such as imagination and creativity for physical humans is complex and troublesome, therefore, attempting to model it in a virtual human is an understandable challenge. Whilst the terms may have some overlap, there does appear to be some consensus that 'imagination' refers to a primarily mental process, whereas 'creativity' refers more to making that imagination manifest in some physical or cultural way (Gaut and Livingstone, 2003), such as sculpture or poetry, or in a more mundane way, such as solving a social, business or technical problem. There is also a sense that imagination precedes, or is closely integrated with, creativity. Boden (1998) refers to these as H-creativity (new to the history of the world) and P-creativity (new just to the psych of the progenitor). However, as an act of imagination or creativity becomes more practiced and exercised, then it does, perhaps, fall more closely into the realm of problem-solving, as discussed in the previous section.

There is a long history of research into creativity by computers – often called computational creativity (Ackerman et al., 2017). Boden (1998) identified three forms of computational creativity (CC):

- Combinational, where the computer 'produces unfamiliar combinations of familiar ideas, and it works by making associations between ideas that were previously only indirectly linked' – such as creating a photo-montage;
- Exploratory, where the computer explores possibilities within a 'culturally accepted style of thinking, or 'conceptual space' – such as a major or minor scale in music; and
- Transformational, where the computer removes or changes the constraints on an accepted conceptual space and then explore this new space.

Computational creativity has been applied across almost every artistic and creative field, for instance:

- BRUTUS, a story-making and telling machine (Bringsjord and Ferrucci, 1999);
- The Painting Fool project (Colton and Wiggins, 2012);

- The Wishful Automatic Spanish Poet (WASP) generator of poetry (Gervás, 2018);
- The current interest in affectively driven algorithmic composition (AAC), which uses computers to generate music with emotional or affective qualities, such as in the work of Williams et al. (2015) and Scirea et al. (2017).

Loughran and O'Neill (2017) analysed submissions to the International Joint Workshops in Computational Creativity (IJWCC) over a period of 12 years and submissions to International Computational Creativity Conferences (ICCC), and identified that across 353 papers the most popular areas for computational creativity were Music, Other (covering a range from Cocktails to Choreography), Imagery, and Language (including Story and Literature), with Games, Humour, Sound and Math further down the list.

There has been a suggestion for a modified Turing Test, where users try to identify which 'cultural artefacts' were created by a computer, and which by a human, and Lamb et al. (2018) situates this within a broader 4Ps evaluation of creativity – Person (or virtual person), Process, Product and Press (namely the audience, media and culture).

However, in all these examples, there is often no claim that the computer is showing 'imagination'; it is following a set of rules or algorithms, or using neural network or machine learning techniques, to create the resultant artefacts. In many ways, this computer creativity is, perhaps, more akin to the aleatory creativity of a musician or author such as John Cage – using i-Ching techniques to randomly create compositions with sounds and words (Blair, 2016). Such works are, perhaps, examples of human creativity without imagination, although of course there must first have been the imagination to try the techniques, and also in framing them, and then curating the result.

It would seem that endowing a virtual human with 'creativity' is probably not a major challenge; the significant challenge is that found in the i-Ching/John Cage example, creating the motivation to be creative in the process, and then the imagination to choose the tools, methods, subject, to frame the piece, and then to act as editor and curator afterwards.

Research into computer imagination seems far less advanced. Indeed, Mahadevan (2018) proposes a new overarching challenge for AI, that of designing 'imagination machines'. Stokes (2016, 259) suggests

> [T]hat imagination is 'non truth-bound' and decoupled from action. This freedom or playfulness of imagination is crucial to the novelty required of creativity, since it allows one to safely 'try out' hypotheses, conceptual combinations, strategies for solution, and so on, without epistemic or behavioral commitment. And because imagination connects in important ways with inferential systems, as well as affective systems, the thoughts it produces can often be integrated with knowledge and skills to formulate an innovative strategy or solution to a problem.

Stokes (2014) also argues that imagination is the key player in the cognitive manipulation required for creativity and highlights the voluntary nature of the imagination and its links to not only cognitive playfulness but also to cognitive work, and the existence of cognitive contagion where affective and other states can bleed across from the imagination to the 'real' self. This idea of imagination as a virtual sandbox could open up interesting opportunities for virtual humans, who could create instances of virtual worlds almost at will in order to unbind and decouple themselves from the truth of the physical world, and have freedom to play and try out ideas. Indeed, this model is very similar to that also proposed for the dream-state with a virtual human (Svensson and Thill, 2013) and the ideas of 'playing away' from identities as discussed in Chapter 10.

META-MANAGEMENT AND SELF-MONITORING

Although most people do not have an explicit sense of their own changes over time, it is only natural for preferences, approaches and beliefs to mature and evolve as they grow and build life experiences. It is also useful if a person can have some self-referential sense of their own state-of-mind, being able to detect when they are tired or stressed or being potentially abusive or discriminatory. A well-developed virtual human could be expected to have similar capabilities, and, indeed, the ability to evolve and adapt over time may be what provides virtual humans with longevity in the same way as it has done for physical humans.

Meta-management and self-monitoring are the processes in a virtual human which monitor what the rest of the virtual human's processes are doing, and then, over time, change them, usually to obtain a better performance or to avoid major issues and failures. Several of the cognitive architecture models to be discussed in Chapter 6 include a meta-management function. For instance, a virtual human may change its own personality as its experiences push it from an introverted personality to a more extroverted one. Another potential role is to act as watch-keeper and step in to change the behaviour if it gets stuck in a loop – for example, overplanning. At its most extreme, this meta activity could result in a virtual human's increased awareness of itself and its actions, potentially manifested through an internal dialogue.

By having a well implemented meta-cognitive function, it should be possible to ensure that virtual humans are more robust and that they can cope with a wider variety of situations without entering into some form of failure mode. Indeed, one of the aims of meta-cognition is to de-risk the other constituent technologies of a virtual human, ensuring that deviant behaviour by any one part does not affect the whole system. As the lower level functions of a virtual human mature and are integrated, the more likely that unexpected consequences of their design and behaviour will become apparent. This will increase the need for the meta-cognitive layer to ensure that the virtual human does not become side-tracked or suffer from analysis paralysis.

CONCLUSION

The development of the different elements that make up a virtual human's mind is far more disparate than the developments of the components of the body and senses considered in Chapter 3. This is reflected in the different drivers for the development of each element. Whilst for the virtual human body the drivers are largely coming from the game and movie industries, there is not a common driver within the different elements of the mind.

Those elements closely linked to sensory systems, such as attention and, to a lesser extent, perception are attracting a lot of interest from commercial and academic robotic and autonomous systems researchers and developers. The work on emotions is being driven more by the games industry and the study of affective computing, and, to a lesser extent, the medical industry in developing empathic care systems. Personality and longer-term motivational systems are less well developed and researched beyond the confines of that needed for NPCs in video games. Virtual humans are still seen as 'tools' to support human activities; giving them their own motivations seems a low priority. The more tactical tasks of decision-making and BDI-type processes are, in contrast, far more developed, as are the very generic capabilities of memory and learning.

Having investigated the elements of the body, senses and mind of a virtual human, Chapter 5 will consider the abilities that a virtual human might have to communicate. Chapter 6 will then draw the different elements together by examining cognitive architectures.

REFERENCES

Ackerman, M., Goel, A., Johnson, C. G., Jordanous, A., León, C., y Pérez, R. P., & Ventura, D. (2017). Teaching computational creativity. In *Proceedings of the 8th International Conference on Computational Creativity* (pp. 9–16). Atlanta.

Adam, C., & Lorini, E. (2014). A BDI emotional reasoning engine for an artificial companion. In: Corchado J. M. et al. (Eds.), *Highlights of Practical Applications of Heterogeneous Multi-Agent Systems. The PAAMS Collection. PAAMS 2014. Communications in Computer and Information Science* (vol. 430). Cham, Switzerland: Springer.

Agüero, C. E., Martín, F., Rubio, L., & Cañas, J. M. (2012). Comparison of smart visual attention mechanisms for humanoid robots. *International Journal of Advanced Robotic Systems, 9*(6), 233.

Aguilar, R. A., de Antonio, A., & Imbert, R. (2007). Emotional Agents with team roles to support human group training. In Pelachaud, C., Martin, J. C., André, E., Chollet, G., Karpouzis, K., & Pelé, D. (Eds.), *Intelligent Virtual Agents. IVA 2007. Lecture Notes in Computer Science*, (pp. 352–353). vol 4722. Berlin, Germany: Springer.

Andriamasinoro, F. (2004). Modeling natural motivations into hybrid artificial agents. In *Fourth International ICSC Symposium on Engineering of Intelligent Systems (EIS 2004).*

Anthony, B., Majid, M. A., & Romli, A. (2017). Application of intelligent agents and case based reasoning techniques for green software development.*Technics Technologies Education Management, 12*(1), 30.

Baddeley, A. (2012). Working memory: Theories, models, and controversies. *Annual Review of Sychology, 63*, 1–29.

Becker-Asano, C. (2014). WASABI for affect simulation in human-computer interaction. In *Proceedings of the International Workshop on Emotion Representations and Modelling for HCI Systems* (pp. 1–10). Istanbul: ACM.

Belbin, R. M. (2012). *Team Roles at Work.* London, UK: Routledge.

Berners-Lee, T., Hendler, J., & Lassila, O. (2001). The semantic web. *Scientific American, 284*(5), 28–37. Available online http://web.cs.miami.edu/home/saminda/csc688/tblSW.pdf.

Blair, M. (2016). Crossroads of boulez and cage: Automatism in music. Available online https://mdsoar.org/bitstream/handle/11603/3751/Verge13_BlairMark.pdf?sequence=1.

Boden, M. A. (1998). Creativity and artificial intelligence. *Artificial Intelligence, 103*(1–2), 347–356.

Bogdanovych, A., Trescak, T., & Simoff, S. (2015). Formalising believability and building believable virtual agents. In Chalup, S. K., Blair, A. D., &

Randall, M. (Eds.), *Artificial Life and Computational Intelligence.* ACALCI 2015. Lecture Notes in Computer Science (vol. 8955, pp. 142–156). Cham, Switzerland: Springer.

Borji, A., Sihite, D. N., & Itti, L. (2012). Modeling the influence of action on spatial attention in visual interactive environments. *Robotics and Automation (ICRA), 2012 IEEE International Conference on* (pp. 444–450). IEEE.

Breazeal, C. (2004). *Designing Sociable Robots.* Cambridge, MA: MIT Press.

Bredeche, N., Montanier, J. M., Liu, W., & Winfield, A. F. (2012). Environment-driven distributed evolutionary adaptation in a population of autonomous robotic agents. *Mathematical and Computer Modelling of Dynamical Systems, 18*(1), 101–129.

Bringsjord, S., & Ferrucci, D. (1999). *Artificial Intelligence and Literary Creativity: Inside the Mind of Brutus, A Storytelling Machine.* London, UK: Psychology Press.

Cattell, R. B., Eber, H. W., & Tatsuoka, M. M. (1970). *Handbook for the 16PF.* Champaign, IL: IPAT.

Cid, F., Moreno, J., Bustos, P., & Núñez, P. (2014). Muecas: A multi-sensor robotic head for affective human robot interaction and imitation. *Sensors, 14*(5), 7711–7737.

Cohen, P. R., & Feigenbaum, E. A. (Eds.), (2014). *The Handbook of Artificial Intelligence* (Vol. 3). Oxford, UK: Butterworth-Heinemann.

Colton, S., & Wiggins, G. A. (2012). Computational creativity: The final frontier? In de Raedt, L., Bessiere, C., Dubois, D., & Doherty, P. (Eds)., *Proceeding ECAI Frontiers* (pp. 21–16). Amsterdam, the Netherlands: IOS Press.

Crowder, J., & Friess, S. (2012). Artificial psychology: The psychology of AI. In *Proceedings of the 3rd Annual International Multi-Conference on Informatics and Cybernetics.* Orlando, FL.

Davis, E., & Morgenstern, L. (2004). Introduction: Progress in formal common sense reasoning. *Artificial Intelligence, 153*(1), 1–12.

Doce, T., Dias, J., Prada, R., & Paiva, A. (2010). Creating individual agents through personality traits. *10th International Conference on Intelligent Virtual Agents* (pp. 257–264). Berlin, Germany: Springer.

Ekman, P. (1989). The argument and evidence about universals in facial expressions of emotion. In H. Wagner & A. Manstead (Eds.), *Handbook of Social Psychophysiology* (pp. 143–164). Chichester, UK: Wiley.

Frith, C., & Frith, U. (2005). Theory of mind. *Current Biology, 15*(17), R644–R645.

Gaut, B., & Livingstone, P. (Eds). (2003) *The Creation of Art: New Essays in Philosophical Aesthetics.* Cambridge, MA: Cambridge University Press.

Gervás, P. (2018). Computer-driven creativity stands at the forefront of artificial intelligence and its potential impact on literary composition. *AC/E Digital Culture Annual Report.: Digital Trends in Culture. Focus: Readers in the Digital Age,* 88.

Goldman, A. I. (2012). Theory of mind. In E. Margolis., R. Samuels., & S. P. Stich (Eds.), *The Oxford Handbook of Philosophy of Cognitive Science.* New York: Oxford University Press.

Hopgood, A. A. (2011). *Intelligent Systems for Engineers and Scientists.* Boca Raton, FL: CRC press.

Izard, C. E. (2013). *Human Emotions.* New York: Springer Science & Business Media.

Jones, N. A., Ross, H., Lynam, T., Perez, P., & Leitch, A. (2011). Mental models: An interdisciplinary synthesis of theory and methods. *Ecology and Society, 16*(1). Available online https://www.ecologyandsociety.org/vol16/iss1/art46/.

Johnson-Laird, P. J. (2005). Mental models and thought. In K. Holyoak & B. Morrison (Eds.), *The Cambridge Handbook of Thinking and Reasoning* (pp. 185–208). Cambridge, MA: Cambridge University Press.

Kanov, M. (2017). *"Sorry, what was your Name Again?": How to Use a Social Robot to Simulate Alzheimer's Disease and Exploring the Effects on its Interlocutors.* Stockholm, Sweden: KTH Royal Institute of Technology.

Kefalas, P., Sakellariou, I., Basakos, D., & Stamatopoulou, I. (2014). A formal approach to model emotional agents behavior in disaster management situations. In Likas A., Blekas K., Kalles D. (Eds.), *Artificial Intelligence: Methods and Applications*, SETN 2014, Lecture Notes in Computer Science (vol. 8445, pp. 237–250). Cham, Switzerland: Springer.

Kelley, T. D. (2014). Robotic dreams: A computational justification for the post-hoc processing of episodic memories. *International Journal of Machine Consciousness, 6*(2), 109–123.

Khalili, A., Auer, S., & Ngonga, Ngomo AC. (2014). Context–Lightweight text analytics using linked data. In: V. Presutti, C. d'Amato, F. Gandon, M. d'Aquin, S. Staab, A. Tordai (Eds.), *The Semantic Web: Trends and Challenges*, ESWC 2014, Lecture Notes in Computer Science (vol. 8465, pp. 628–643). Cham, Switzerland: Springer.

Kiesel, D. (2005). A brief introduction to neural networks (ZETA2-EN). Available online http://www.dkriesel.com/_media/science/neuronalenetze-en-zeta2-2col-dkrieselcom.pdf.

Lamb, C., Brown, D. G., & Clarke, C. L. (2018). Evaluating computational creativity: An interdisciplinary tutorial. *ACM Computing Surveys(CSUR), 51*(2), 28.

Lilly, J. (2010). *Programming the Human Biocomputer.* Berkeley, CA: Ronin Publishing.

Lim, M. Y., Dias, J., Aylett, R., & Paiva, A. (2012). Creating adaptive affective autonomous NPCs. *Autonomous Agents and Multi-Agent Systems, 24*(2), 287–311.

Lorenz, K., & Leyhausen, P. (1973). *Motivation of Human and Animal Behavior; An Ethological View.* New York: van Nostrand Reinhold Company.

Loughran, R., & O'Neill, M. (2017). Application domains considered in computational creativity. In A. Goel., A. Jordanous, & A. Pease (Eds.), *Proceedings of the Eighth International Conference on Computational Creativity ICCC 2017.* Atlanta, Georgia, June 19–23.

Mahadevan, S. (2018). Imagination machines: A new challenge for artificial intelligence. Available online https://people.cs.umass.edu/~mahadeva/papers/aaai2018-imagination.pdf.

Maslow, A. (1954). *Toward a Psychology of Being*, 3rd ed. London, UK: Wiley & Sons.

Mataric, M. J. (1994). Reward functions for accelerated learning. *Machine Learning Proceedings*, 181–189.

Mautner, J., Makmal, A., Manzano, D., Tiersch, M., & Briegel, H. J. (2014). Projective simulation for classical learning agents: A comprehensive investigation. *New Generation Computing, 33*(1), 69–114.

McCormick, J., Vincs, K., Nahavandi, S., Creighton, D., & Hutchison, S. (2014). Teaching a digital performing agent: Artificial neural network and hidden arkov model for recognising and performing dance movement. In *Proceedings of the 2014 International Workshop on Movement and Computing* (p. 70). New York: ACM.

McRorie, M., Sneddon, I., McKeown, G., Bevacqua, E., De Sevin, E., & Pelachaud, C. (2012). Evaluation of four designed virtual agent personalities. *Affective Computing, IEEE Transactions on, 3*(3), 311–322.

Minton, S., Carbonell, J. G., Knoblock, C. A., Kuokka, D. R., Etzioni, O., & Gil, Y. (1989). Explanation-based learning: A problem solving perspective. *Artificial Intelligence, 40*(1–3), 63–118.

Moore, G. T., and R. G. Golledge. 1976. Environmental knowing: Concepts and theories. In G. T. Moore & R. G. Golledge (Eds.), *Environmental Knowing: Theories, Research and Methods* (pp. 3–24). Stroudsburg, PA: Dowden Hutchinson and Ross Inc.

Ortony, A., Clore, G. L., & Collins, A. (1990). *The Cognitive Structure of Emotions.* Cambridge, MA: Cambridge University Press.

Parasuraman, R., Sheridan, T. B., & Wickens, C. D. (2000). A model for types and levels of human interaction with automation. *IEEE Transactions on systems, man, and cybernetics-Part A: Systems and Humans, 30*(3), 286–297.

Pease, A., Chaudhri, V., Lehmann, F., & Farquhar, A. (2000). Practical knowledge representation and the DARPA high performance knowledge bases project. In *KR-2000: Proceedings of the Conference on Knowledge Representation and Reasoning* (pp. 717–724). Breckenridge, CO.

Pérez-Pinillos D., Fernández S., Borrajo D. (2013). Modeling motivations, personality traits and emotional states in deliberative agents based on automated planning. In J. Filipe, A. Fred (Eds.), *Agents and Artificial Intelligence*, ICAART 2011, Communications in Computer and Information Science (vol. 271, pp. 146–160). Berlin, Germany: Springer.

Perner, J. (1991). *Understanding the Representational Mind.* Cambridge, MA: The MIT Press.

Pomerol, J. C. (1997). Artificial intelligence and human decision making. *European Journal of Operational Research, 99*(1), 3–25.

Prinz, J. J. (2006). Is emotion a form of perception? *Canadian Journal of Philosophy,36*(supl), 137–160.

Puică, M. A., & Florea, A. M. (2013). Emotional belief-desire-intention agent model: Previous work and proposed architecture. *International Journal of Advanced Research in Artificial Intelligence, 2*(2), 1–8.

Rahwan, T., Rahwan, T., Rahwan, I., & Ashri, R. (2004). Agent-based support for mobile users using agentSpeak (L). In P. Giorgini, B. Henderson-Sellers

& M. Winikoff. (Eds.), *Agent-Oriented Information Systems (AOIS 2003): Revised Selected Papers 45–60*. Berlin, Germany: Springer.

Rao, A. S., & Georgeff, M. P. (1991). Modeling rational agents within a BDI-architecture. In: R. Fikes and E. Sandewall (Eds.), *Proceedings of Knowledge Representation and Reasoning (KR&R-91)* (pp. 473–484). Burlington, MA: Morgan Kaufmann.

Rozo, L., Jiménez, P., & Torras, C. (2013). A robot learning from demonstration framework to perform force-based manipulation tasks. *Intelligent Service Robotics, 6*(1), 33–51.

Salgado, R., Bellas, F., Caamano, P., Santos-Diez, B., & Duro, R. J. (2012). A procedural long term memory for cognitive robotics. In *Evolving and Adaptive Intelligent Systems (EAIS), 2012 IEEE Conference on* (pp. 57–62). IEEE.

Salichs, M. A., & Malfaz, M. (2012). A new approach to modeling emotions and their use on a decision-making system for artificial agents. *Affective Computing, IEEE Transactions on, 3*(1), 56–68.

Sansonnet, J. P., & Bouchet, F. (2013) A framework covering the influence of FFM/NEO PI-R traits over the dialogical process of rational agents. In J. Filipe & A. Fred (Eds.) *Agents and Artificial Intelligence, ICAART 2013, Communications in Computer and Information Science* (vol. 449, pp. 62–79). Berlin, Germany: Springer.

Schank, R. C. (1990). *Tell Me A Story: A New Look at Real and Artificial Memory*. New York: Charles Scribner's Sons.

Scherer, K. R., Bänziger, T., & Roesch, E. (2010). *A Blueprint for Affective Computing: A Sourcebook and Manual*. Oxford, UK: Oxford University Press.

Scirea, M., Togelius, J., Eklund, P., & Risi, S. (2017). Affective evolutionary music composition with meta compose. *Genetic Programming and Evolvable Machines, 18*(4), 433–465.

Si, M. (2015). Should I stop thinking about it: A computational exploration of reappraisal based emotion regulation. *Advances in Human-Computer Interaction*, 5.

Simons, D. J., & Chabris, C. F. (1999). Gorillas in our midst: Sustained inattentional blindness for dynamic events. *Perception-London, 28*(9), 1059–1074.

Singh, P. (2002) 'The open mind common sense project.' Available online http://www.kurzweilai.net/.

Slater, S., & Burden, D. (2009). Emotionally responsive robotic avatars as characters in virtual worlds. In *Games and Virtual Worlds for Serious Applications, 2009. VS-GAMES'09. Conference* (pp. 12–19). IEEE.

Stachowicz, D., & Kruijff, G. J. M. (2012). Episodic-like memory for cognitive robots. *Autonomous Mental Development, IEEE Transactions on, 4*(1), 1–16.

Stokes, D. (2014). The role of imagination in creativity. In E. S. Paul & S. B. Kaufman (Eds) *The philosophy of Creativity: New essays* (pp. 157–184). Oxford, UK: Oxford Scholarship Online.

Stokes, D. (2016) Imagination and creativity, in A. Kind (Ed.), *The Routledge Handbook of the Philosophy of Imagination*. London, UK: Routledge, 247–261.

Sutton, S. and Barto, A. G. (1998). *Reinforcement Learning: An Introduction*. Cambridge, MA: MIT press.

Svensson, H., & Thill, S. (2013) Should robots dream of electric sheep? *Adaptive Behavior, 21*(4), 222–238.

Vernon, D., Beetz, M., & Sandini G (2015). Prospection in cognition: The case for joint episodic-procedural memory in cognitive robotics. *Frontiers in Robotics & AI, 2*, 19

W3C. (2014). RDF 1.1 concepts and abstract syntax. W3C recommendation. Available online http://www.w3.org/TR/2014/REC-rdf11-concepts-20140225/.

Williams, D., Kirke, A., Miranda, E. R., Roesch, E., Daly, I., & Nasuto, S. (2015). Investigating affect in algorithmic composition systems. *Psychology of Music, 43*(6), 831–854.

Wilson, I. (2000). The artificial emotion engine, driving emotional behavior. In *AAAI Spring Symposium on Artificial Intelligence and Interactive Entertainment* (pp. 20–22). Palo Alto, CA. Available online http://www.aaai.org/Papers/Symposia/Spring/2000/SS-00-02/SS00-02-015.pdf.

CHAPTER 5

Communications

INTRODUCTION

Chapters 3 and 4 have explored the ways in which the body, mind and senses of a virtual human can be modelled. This chapter explores how a virtual human can communicate using both non-language modalities, such as gesture, and language-based modalities, such as text messaging and speech. In the latter, the key techniques are those of natural language understanding and natural language generation. Whilst it is natural language understanding which seems to have been the focus of much of the research work over the last few decades, natural language generation is nowhere near as well researched or developed, with the majority of systems using some form of templating approach. Having considered the use of a language to communicate *between* entities, the chapter will conclude with a consideration of the importance of language to the internal narrative of the virtual human.

COMMUNICATIONS: NON-LANGUAGE MODALITIES

Whilst Mehrabian's 7-38-55 Rule, that 55% of our total communication is delivered by body language, 38% by vocal signals and only 7% by words, may be justifiably challenged (Lapakko, 1997), it does highlight the perceived importance of non-verbal communications in human interaction. A virtual human lacking any non-verbal communication will seem very robotic indeed – even C3PO manages some expressive hand waving.

Kelly (2001) identifies the importance of non-verbal behaviours in everyday face-to-face interactions, and how those behaviours can both clarify what is being said and reveal intentions as well.

Cassell (2001) describes how both speech and non-speech modalities can fulfil the propositional (content) and interactional (signalling) functions within a dialogue, and how that fits within a model of conversation: Function, Modality, Behaviour and Timing (FMBT, pronounced fem-bot). The FMBT model has been implemented by Cassell (2001, 75) in an embodied conversational agent (ECA) called REA:

> In the implementation of REA, we have attended to both propositional and interactional components of the conversational model, and all the modalities at REA's disposal (currently, speech with intonation, hand gesture, eye gaze, head movement, body posture) are available to express these functions. REA's current repertoire of interactional functions includes: acknowledgement of a user's presence, feedback, and turn taking.

Since virtual humans have the ability to both trigger and read gestures and facial expressions, then they should be just as capable of engaging in non-language-based communication as physical humans.

COMMUNICATIONS: LANGUAGE-BASED MODALITIES

Despite Mehrabian's rule, most people recognise that for practical purposes, language-based modalities (such as speech and text) are the primary means of communications for humans, and so they should also be the primary means for virtual humans. This section will examine speech recognition and speech generation technologies, followed by a discussion about natural language communications and the related areas of natural language generation, storytelling and argumentation.

Speech Recognition

Speech recognition is primarily regarded as the ability for a computer to understand spoken words, although it can also include the ability of the computer to recognise emotion through speech, and also voice recognition – the ability to identify individuals from their speech patterns. The growth in consumer mobile systems and Interactive Voice Response (IVR) systems for customer service applications has seen significant work on speech recognition over the last few decades, but there can still be a sizeable gulf between how well users expect it to work and how well it works in practice (particularly, in less than ideal, noisy environments). The challenge of speech recognition is summed up by the notion that, traditionally, a system could be developed to reliably understand almost all words from one person, or a small selection of words from almost everyone, but developing a system to reliably understand almost all words from almost everyone has proven to be a major challenge. For example, IVR systems, as used in telephone services, are able to cope with a wide range of voices since they have a language model (with supporting vocabulary and grammar), which is constrained to just the task at hand (for example, booking a cinema ticket), so the classifier has a limited number of options that it needs to choose between.

Cutajar et al. (2013) provides a useful review of automated speech recognition (ASR) techniques. She describes how ASR takes advantage of the momentary (~10ms) gaps in human speech between the phonemes (sound units) that make up words, with English only consisting of 42 distinct phonemes. Cutajar describes the typical ASR process.

- **Pre-processing** – for example, removing noise and silence.
- **Feature extraction** – identification of the sonic features that will help to identify the phonemes.
- **Language model** – to inform the classification process, which contains information on semantics and syntax. This can help with classification if the speech is expected to conform to a particular grammar.
- **Classification** – the application of the language model to the features to identify the phonemes being said, often using a hybrid approach using generative and discriminative approaches.

It is partly because of the link between speech recognition and the language model that speech recognition is often shown bridging the 'body' and 'brain' components of the virtual human architecture. This is because the more knowledge that an ASR system has of what might be being said (given spatial, social and discursive context) the easier it is to identify the phonemes being said (although this does, of course, introduce the chance of misperception).

Much recent research has been exploring the use of neural network and machine learning approaches to speech recognition.

For instance, Hinton states that 'deep neural networks (DNNs) … have been shown to outperform [existing approaches] on a variety of speech recognition benchmarks, sometimes by a large margin' (Hinton et al., 2012). Hinton reports that on a test using mobile-phone based Bing search engine queries, DNN techniques achieved a 71.7% accuracy rate (as against 63.8% for traditional approaches), and on a switchboard test (where there was a more limited vocabulary), an 81.5% accuracy rate (as against 72.6%). McMillan et al. (2015) suggests similar accuracy rates in commercial systems such as Siri and the Google Speech API. According to Johnson et al (2014), looking at speech recognition (SR) in healthcare accuracy rates can be as high as 98.5% for SR, as against a highest of 99.6% for human transcription – domain-specific vocabularies no doubt helping the task.

The approach to emotion detection in speech is largely similar to that of speech recognition, but with an interest in different features and classifiers. Koolagudi and Rao (2012) identify three important speech features for emotion detection, in increasing order of use: excitation source, vocal tract system, and prosodic features. In terms of accuracy, speech emotion recognition systems can achieve accuracy rates of up to around 95% (Pan, 2012) on a three-emotion state (happy, sad, neutral) model.

A summary of voice print recognition work – identifying the speaker from the voice – is provided by Zheng et al. (2014) who reports error rates of anywhere between 1% and 20%, depending on conditions, and less than 1% for optimised systems. Finally, there is also active work on getting ASR to work across multiple languages (for example, Besacier et al., 2014) with Siri, for instance, working across (only) eight languages (in 2012) when there are 6,909 known languages.

Enabling speech recognition for a virtual human (or indeed many other computer system) has a number of benefits, such as being able to:

- Talk to the system or virtual human in the same way that one would talk to a physical human, with, ideally, no special commands needed, making access to information and knowledge potentially as easy as talking to a person;
- Use or talk to the system/virtual human when hands (and possible attention) are in use elsewhere (for example, whilst driving); and
- Talk to a virtual character in a training simulation increasing the immersion, the reality of the training and the range of skills that can be trained and practiced.

Given the interest of companies such as Apple (Siri), Google (Home), Amazon (Alexa) and Microsoft (Cortana), and the increasing prevalence of their virtual personal assistants (especially on mobile and in the home), speech recognition technology is being rapidly advanced, although there still seems to be a gap between expectations, laboratory assessments and reality.

The renewal of interest in virtual reality is also emerging as a new driver for speech recognition as voice is usually a better user interface than text within such systems given the fact that the user can't usually see a keyboard, is often mobile, and placing a 2D text display on-screen within a 3D world can cause additional focusing issues.

Speech Generation (Text to Speech)

The most natural way of communicating with a virtual human is through speech. Whilst speech recognition (the virtual human understanding speech) is still challenging, text-to-speech (TTS) technology (enabling the virtual human to speak) is far more mature, if not particularly perfect. The text that the virtual human needs to say is generated by some form of natural language system (see below) and the TTS engine converts this into audio.

Taylor (2009) provides a useful introduction to text-to-speech synthesis. He identifies unit selection synthesis as being the dominant TTS system since the early 90s. In this method, a corpus of audio recordings is split into a large number of multiple di-phone elements (a di-phone being the second half of one phonetic element plus the first half of the following phonetic element). The software then analyses the text and chooses the right sequence of concatenated elements in order to produce the speech. However, the resultant voices are often missing the prosodic elements of 'sound duration, loudness, emphasis, pitch, pauses, and other so-called suprasegmental aspects of speech' (Taylor, 2009).

To avoid the need for huge databases of speech fragments in different styles, recent effort has focused on Statistical Parametric Speech Synthesis (SPSS) based on hidden Markov models which synthetically generate the speech by (in part) modelling the behaviour – or at least the output – of the human larynx, but 'the naturalness of the synthesized speech from SPSS is still as not as good as that of the best samples from concatenative speech synthesizers' (Zen, 2014).

As an example of the work on expression, Black describes an SPSS-based system using Articulatory Features, 'non-segmental features of speech such as nasality, aspiration, voicing' (Black et al., 2012). In tests, their system achieved 'an average classification accuracy of 60% on the four-emotion problem (anger, sadness, neutral, happiness), with most confusion appearing between happiness, neutral, and sadness'. With 15 emotions, the average fell to 12%, and this compares to a reference of 41% accuracy on a database of seven emotions using recorded real speech.

As well as creating more natural speech, another focus has been on creating a TTS system across a wider range of languages. Most major languages now have TTS systems available, including Arabic, Mandarin Chinese and Hindi, and research is now taking place on creating TTS systems for other languages, such as Kannada, Tamil and Gujarati. Available TTS systems include AT&T Natural Voices, Nuance (owners of Dragon Naturally Speaking), Neospeech, Oddcast (with an easy-to-use web-based system called Sitepal), Festival and eSpeak (both free systems).

Another area of interest is in creating TTS systems that reflect specific individuals, sometimes called voice-fonts. Again, SPSS systems offer more scope for this than concatenated unit-selection systems, as the latter would require each subject to read through the training corpus (1,800 utterances in one system). One area that is driving research in this area is that of providing speech systems for those with degenerative diseases, such as motor neurone disease and Parkinson's disease. Yamagishi et al. (2012) describe how SPSS has been used to create personalized voice fonts from only 5–7 minutes of recorded speech, and also how a voice-bank is being built not only of patients whose voice is failing, but also of relatives and similar sounding voice donors. Kisner (2018) reports how a similar commercial system – VocalID – allows users to create their own voice-fonts, or have their voices reconstructed from existing recordings of their own speech. Similar systems are Modeltalker (https://www.modeltalker.org/), which also allows you to create your own voice font (needing 1,600 sentences), and Lyrebird (https://lyrebird.ai/), which works from only one minute of audio – although the audio quality from such a small sample is barely usable or identifiable.

The Speech Synthesis Markup Language, SSML (Taylor and Isard 1997), is a commonly used platform independent interface standard for speech synthesis systems, which can be used to mark up the prosody elements of speech to complement the words in the text. It is a formal Recommendation of the World Wide Web Consortium's voice browser working group.

Speech is such a natural interface that having virtual humans able to converse, even if only one way, in speech makes the experience more engaging, improving training and learning, and increasing the emotional bond between user and computer, which can result in higher rates of disclosure and trust.

NATURAL LANGUAGE UNDERSTANDING AND COMMUNICATION

Natural language is how physical humans most commonly interact, so extending this capability to virtual humans is essential. The provision of such an ability can be split into two elements, the conversion to and from the communication medium (be it speech or sign language) to a text (or other symbolic) representation of language (such as the content of what is said), and the handling of the symbolic language itself. The latter part divides into three:

- Understanding the language to work out what the user wants or is saying,
- Working out what content the reply should contain, and then
- Generating the language to impart that content.

Within this section, it will be assumed that the natural language conversation is being carried out by text. If being carried out by speech, then the speech recognition and text-to-speech systems described above can be used to translate to and from text, although, as noted above, a close linkage between the speech recognition and natural language understanding elements can significantly enhance performance.

Bibauw et al. (2015) provides a useful overview and taxonomy of natural language systems and approaches within the context of language learning, as illustrated in Figure 5.1.

It is useful to consider three broad approaches to natural language understanding systems: pattern matching, grammatical parsers and machine learning.

Artificial Intelligence Markup Language (AIML) (Wallace, 2003) is probably the most common pattern matching natural language system. Although it started off as a hobbyist system, it is now used commercially (for example, Pandorabots); it has also been behind many of the finalists and winners in the annual Loebner Prize (an implementation of the Turing Test) (Bradeško and Mladenić, 2012). AIML simply compares the user input, character by character, to a set of patterns in its database, and each pattern is linked to one or more responses that are used if the pattern is the best match for inputted text. The patterns can include wild cards and other features, so a one-for-one match is not needed, but it is still a relatively brute-force approach. Whilst AIML

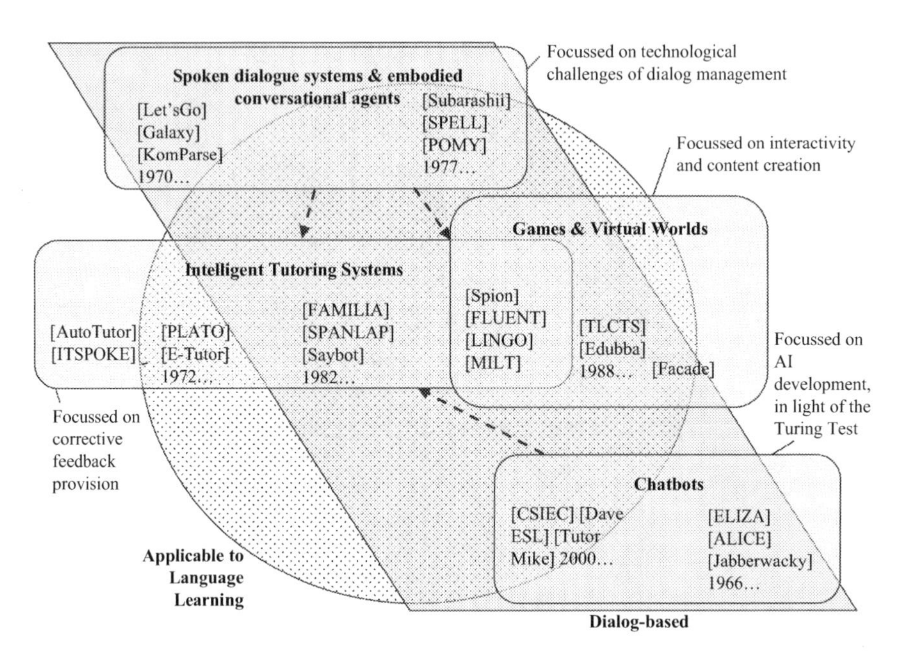

FIGURE 5.1 Research on dialogue systems for language learning mapped against four different disciplines. (After Bibauw, S. et al., *Conversational Agents for Language Learning: Sate of the Art and Avenues for Research on Task-based Agents,* CALICO edition, Boulder, CO, 2015.)

makes no pretence to 'understand' what the user is saying, the performance, if well authored, can be very good. However, it does suffer from the need to write more and more cases as the breadth of knowledge increases, making it difficult to scale up.

A subset of pattern matching systems are rules-based systems (for example, Bruce Wilcox's Chatscript – http://chatscript.sourceforge.net/). This has a set of rules which fire when certain input conditions are met, and, by default, rules are only fired once to avoid repetition, and the rule response can include a sequence of tests and actions defined by an internal scripting language. Daden (Savin-Baden et al., 2013) took a similar approach in developing a chatbot for use with undergraduate students focusing on health-related topics of discussion. Advanced pattern-match systems (including Chatscript and the latest implementations of AIML) offer further power and flexibility by linking into external APIs (including web services), triples-based semantic databases and grammatical parsers. As such, they represent a hybrid approach, which offers a good short- to medium-term solution to creating powerful chatbots (for example, Jaffe et al., 2015 and Savin-Baden et al, 2013).

In the past, much academic research into natural language understanding was focused around grammatical parsers (for example, Nolan, 2014). These break down the user input into parts of speech (verb, noun, adjective, etc.) in order to try and identify the 'meaning' of what the user is saying, and then generate a reply accordingly. They can be readily linked to semantic knowledge bases to identify the information needed for the response – but face the additional challenge of then having to generate the natural language response as there are no hardcoded responses as in AIML or Chatscript. An additional problem of the semantic approach is that the agent has to be taught to understand how a language works first before it can be taught the specifics of a task. Interestingly, most of the recent papers identified on natural language interfaces (for example, Khanna et al., 2015) tended to be using AIML as that provides a faster route to creating a natural language interface to investigate specific issues. However, there is still a lot of interest in semantic parsing in relation to extracting the knowledge from written texts.

Machine Learning

Much of the most recent activity has been around machine learning techniques, where the chatbot uses a large corpus of previous conversations and then matches the input with the input to a previous conversation, and uses its related output (Vinyals and Le, 2015, from Google's AI team). Such systems depend on a large corpus of previous conversations or information to mine, and on the quality of the machine learning techniques (for example, neural nets) used to find the best match for the input. As such, they are not useful at focused tasks (for example, providing support) unless there is already a large corpus of information to work from, and their linguistic style may vary, as each response may well have come from a different person.

Conversation Management

An associated area is conversation management, moving the bot away from simple question-answer dialogues into a more human-like exchange where the 'lead' in the conversation changes, probably several times, during the conversation. This is significantly informed by the work in the field of conversation and discourse analysis between humans, looking at areas such as turn-taking and speech-acts (Sidnell, 2011). This also enables the virtual human to start engaging in argument, and the study of argumentation within the context of virtual humans is a significant but as yet underdeveloped area (for example, Wei and Prakken, 2017 and Thimm et al., 2016). As the balance in conversation shifts from the human to the virtual human, the latter starts to assume the role of anecdote-teller, or even storyteller, and a whole new set of conversational norms start to apply, and this is, again, an area of intriguing research in virtual humans (for example, Adam and Cavedon, 2013).

The type of discourse between the user and virtual human has an impact on the relationships formed with the virtual human. For example, simpler dialogues based on pattern recognition may be better suited for roles where the user queries a database of factual information, whereas, more advanced semantic-based methods of understanding and creating natural language may provide more adept virtual humans for roles focused on guidance and mentoring. However, it should be noted that the systems described above can only provide virtual humans with appropriate contextually relevant responses through increasingly sophisticated categorisation of user input. Other factors relating to how the virtual human is perceived and relationships between the human user and virtual human are equally important.

Uses and Future Developments

The sophistication of the natural language understanding and the types of conversation generated can have an impact on the roles in which the virtual human can be deployed.

Being able to allow non-player characters to interact with users in a natural way significantly enhances the immersion and range of tasks and training that can be simulated. However, it should be noted that many training systems (for example, Virtual Cultural Awareness Trainer from Alelo – https://www.alelo.com/virtual-cultural-awareness-trainer-vcat/ and the Communicate! System described by Lala et al., 2017) use a pick-list of things for the user to say, and for which there are pre-programmed responses, rather than a full natural language system. The dangers of a more open-ended conversational bot, driven by user-sourced content are shown by Microsoft's Tay chatbot, which was shut down after just 16 hours when it began to generate 'racist and sexist Twitter messages' and 'went on an embarrassing tirade, likening feminism to cancer and suggesting the Holocaust did not happen'. (The Guardian, 2016)

As with speech recognition and generation, the mobile phone and home assistant markets are probably driving the development of natural language systems. Many of the 'bots' on text-chat systems, such as websites and Facebook Messenger, are still using structured dialogues more than pure natural language. Natural language conversations are still a major challenge, as shown by Facebook's purchase of chatbot start-up Wit.ai in 2015, whose staff ended up manning the phones and providing a mechanical Turk-style human interface for users when the software failed to deliver (Gee, 2018). In addition to the commercial and purely academic research, natural language is also an area in which hobbyists (particularly those involved in the Loebner prize), computer game developers and virtual world developers are also all active since the barrier to entry is pleasantly low.

As natural language systems become more complex, they become far less deterministic, so whilst conversations may be more natural, there is more chance of the virtual human saying things which its programmers did not expect it to say. This may range from the inconsequential to the insulting, wrong or even dangerous. Whilst most natural language research is into European-style languages, work on Arabic and other languages is far less developed.

With these limitations, and in a narrow domain, natural language conversation systems can work well now. Covert Turing Tests, where judges and hidden humans do not know a test is taking place and so behave completely naturally, are already being passed (Gilbert and Forney, 2015 and Burden et al., 2016). In 2016, a chatbot named Eugene Goostman fooled 10 out of 30 judges into believing that it was human in a version of the Turing Test staged by Reading University at the Royal Society – and was declared as having achieved the 30% threshold set by Turing (Warwick and Shah, 2016). Although a less scientific example, when the Ashley Maddison sex-dating website was hacked, 70,000 female accounts were found to be run by 'fembots' (Cockayne et al., 2017), compared to around 12,000 real women (Newitz, 2015), and engaging with around 11 million men! The Loebner Prize, an annual, relatively strict implementation of the Turing Test (Abdul-Kader, 2015), continues to challenge chatbot authors into creating believably human conversations.

NATURAL LANGUAGE GENERATION

Whilst natural language understanding is complex, natural language generation (NLG) appears to be a possibly more complex but less researched area. Much of the existing research is linked to natural language presentation of data, rather than more narrative or conversational forms (for example, Ramos-Soto et al., 2015). In all but the grammatical parser approach described above, the systems include the natural language responses, either as whole texts or as templates or rules, rather than having to create the responses word-by-word. As an example of the difference, given the question 'What time is the next train to Birmingham?' a virtual human might retrieve a time of 1800 from a timetable system, and also store the key words 'next', 'train' and 'Birmingham'. If using a template approach, it would have the template 'The [specifier] [transportmode] to [destination] is at [time]', and then fill in the gaps to give 'The next train *to* Birmingham *is at* 1800'. If using a grammatical approach, then the virtual human would just have the words 'next', 'train' 'Birmingham' and '1800' and need to build a sentence from them.

The template approach makes natural language systems far easier to implement. The chatbot is just given the templates it needs, whereas a grammar-based system might need to be taught how to use the whole of English (or another language) first, before being taught the specific expressions it needs to handle. Templates, being human-authored, also provide a far more 'human' output, than current grammar-based systems. However, the result is still far from ideal, as Wen challenges '[d]espite its robustness and adequacy, the frequent repetition of identical, rather stilted, output forms make talking to a rule-based generator rather tedious' (Wen, 2015). Reiter and Dale (1997) refer to the template/rule approach somewhat dismissively as 'mail-merge'. Others, though, defend the approach. For instance, Deemter et al. (2005, 22) argues:

> [T]hat systems that call themselves template based can, in principle, perform all NLG tasks in a linguistically well-founded way and that more and more actually implemented systems of this kind deviate dramatically from the stereotypical systems that are often associated with the term template.

In terms of alternative approaches, Reiter and Dale (1997) describe the steps within a 'corpus-based' NLG system.

1. Content determination (what needs to be communicated).
2. Discourse planning (how each bit needs to be said – which may include micro-templates).
3. Sentence aggregation (how all the bits will fit together).
4. Lexicalization (the exact choice of words to reflect the expressed relations and concepts, some of which may be hard-coded).
5. Referring expression generation (the exact choice of words to reflect the entities referred to).
6. Linguistic realization (a tidy up to ensure that rules of grammar are being followed).

Note that 'corpus-based' in Reiter's case only refers to collecting and using a corpus of existing text in order to inform the creation of the rest of the system, not to using it as an integral part of the system.

An alternative approach is the stochastic language generation system described by Oh (2000). This takes a corpus of sentences used in real life by people and related to the domain of the chatbot. It then breaks this corpus into N-grams (i.e., collections of small numbers of words, including slots for data). When it needs to build a response, it randomly (stochastically) combines these N-grams into possible sentences/replies, then scores them against a heuristic (for example, length, includes correct slots) to find the best response. A more typical modern approach using neural networks is taken by Wen (2015). Wen's Long Short-Term Memory (LSTM) language generator works from a training corpus as with Oh, but working with speech acts and other heuristics to constrain the output during generation rather than afterwards.

It should be noted that all three examples (Reiter and Dale, 1997; Oh, 2000 and Wen, 2015) require reasonable sizes of corpus to work from (for example, Wen used 1,000 dialogues just to ask about San Francisco restaurants), and so extending them to the breadth of language capability needed by a fully-realized virtual human would be a significant challenge, and it may be that an alternative approach is required, based more closely on human approaches.

INTERNAL DIALOGUE

Whilst a person may be involved in hundreds or even thousands of dialogues every day with a variety of different people (and even a few systems), the amount of language which they generate or parse in these exchanges is probably overshadowed by the amount of internal dialogue that they are having with themselves as they debate how they slept, what to have for breakfast, think about things on the way to work and then manage all the challenges and events of the rest of the day.

As a result, there is a vital role for language in the internal dialogue that all humans have, and which should be expected to be present in virtual humans. Clancey (1997) describes the importance of this internal use of language as a cognitive aid, in that it both enables a person to express thoughts and ideas, but also to change those same thoughts and ideas internally. Gazzaniga (2000) discussing his research on split-brain patients and the concept of a 'narrative-self', says that:

> The same split-brain research that exposed startling differences between the two hemispheres revealed that the human left hemisphere harbors our interpreter. Its job is to interpret our responses—cognitive or emotional—to what we encounter in our environment. The interpreter sustains a running narrative of our actions, emotions, thoughts, and dreams. The interpreter is the glue that keeps our story unified and creates our sense of being a coherent, rational agent. To our bag of individual instincts it brings theories about our life. These narratives of our past behavior seep into our awareness and give us an autobiography

CONCLUSION

This chapter has highlighted the importance of non-language modalities in communication, and how these can be utilised by virtual humans. The chapter has discussed the growing capabilities of artificial natural language understanding, speech recognition and the emerging drivers of commercial chatbots, and mobile and domestic virtual assistants. Whilst it may seem that progress towards a comprehensive and sustained passing of the Turing Test is slow, the signs are there that within a less combative environment a well programmed chatbot operating within a reasonably restricted domain can already fool enough humans to make an impact, and that the capability of chatbot systems is only going to increase. Given current rates of improvement, it would seem reasonable for chatbots to regularly pass covert Turing Tests in the next five years, and for sustained success at the standard Turing Test within 5–10 years.

Having examined all the component parts of a virtual human, the next chapter will look at how cognitive architectures pull all these elements together into a complete virtual human.

REFERENCES

Abdul-Kader, S. A., & Woods, J. (2015). Survey on chatbot design techniques in speech conversation systems. *International Journal of Advanced Computer Science and Applications*, 6(7), 72–80.

Adam, C., & Cavedon, L. (2013). A companion robot that can tell stories. In R. Aylett, B., Krenn., C. Pelachaud & H. Shimodaira (Eds.), *Proceedings of Intelligent Virtual Agents*. 13th International Conference, IVA. Heidelberg, Germany: Springer.

Besacier, L., Barnard, E., Karpov, A., & Schultz, T. (2014). Automatic speech recognition for under-resourced languages: A survey. *Speech Communication, 56*, 85–100.

Bibauw, S., François, T., & Desmet, P. (2015). *Conversational Agents for Language Learning: Sate of the Art and Avenues for Research on Task-based Agents*. CALICO edition. Boulder, CO. Available online https://lirias.kuleuven.be/bitstream/123456789/499442/1/2015.05.28+Dialogue+systems+for+language+learning.pdf.

Black, A. W., Bunnell, H. T., Dou, Y., Muthukumar, P. K., Metze, F., Perry, D., Polzehl, T., Prahallad, K., Steidl, S., & Vaughn, C. (2012). Articulatory features for expressive speech synthesis. In *Acoustics, Speech and Signal Processing (ICASSP), 2012 IEEE International Conference on* (pp. 4005–4008). IEEE.

Bradeško, L., & Mladenić, D. (2012). A survey of chatbot systems through a loebner prize competition. In *Proceedings of Slovenian Language Technologies Society Eighth Conference of Language Technologies* (pp. 34–37).

Burden, D. J. H., Savin-Baden, M., & Bhakta, R. (2016). Covert implementations of the turing test: A more level playing field? In *International Conference on Innovative Techniques and Applications of Artificial Intelligence* (pp. 195–207). Cham, Switzerland: Springer.

Cassell, J. (2001). Embodied conversational agents: Representation and intelligence in user interfaces. *AI magazine, 22*(4), 67.

Clancey, W. J. (1997). *Situated Cognition: On Human Knowledge and Computer Representations*. Cambridge, MA: Cambridge University Press.

Cockayne, D., Leszczynski, A., & Zook, M. (2017). # HotForBots: Sex, the non-human and digitally mediated spaces of intimate encounter. *Environment and Planning D: Society and Space, 35*(6), 1115–1133.

Cutajar, M., Gatt, E., Grech, I., Casha, O., & Micallef, J. (2013). Comparative study of automatic speech recognition techniques. *Signal Processing, IET, 7*(1), 25–46.

De Mantaras, R. L., & Arcos, J. L. (2002). AI and music: From composition to expressive performance. *AI magazine, 23*(3), 43.

Deemter, K. V., Theune, M., & Krahmer, E. (2005). Real versus template-based natural language generation: A false opposition? *Computational Linguistics, 31*(1), 15–24.

Ekenel, H. K., Stallkamp, J., Gao, H., Fischer, M., & Stiefelhagen, R. (2007). Face recognition for smart interactions. In *Multimedia and Expo, 2007 IEEE International Conference on* (pp. 1007–1010). IEEE.

Gazzaniga, M. S. (2000). Cerebral specialization and interhemispheric communication: Does the corpus callosum enable the human condition? *Brain, 123*(7), 1293–1326.

Gee, S. (2018). Facebook closes M, Its virtual assistant. i-programmer.info. Available online http://www.i-programmer.info/news/105-artificial-intelligence/11448-facebook-closes-virtual-assistant.html.

Gilbert, R. L., & Forney, A. (2015). Can avatars pass the Turing test? Intelligent agent perception in a 3D virtual environment. *International Journal of Human-Computer Studies, 73*, 30–36.

Hinton, G., Deng, L., Yu, D., Dahl, G. E., Mohamed, A. R., Jaitly, N., … Kingsbury, B. (2012). Deep neural networks for acoustic modeling in speech recognition: The shared views of four research groups. *Signal Processing Magazine, IEEE, 29*(6), 82–97.

Jaffe, E., White, M., Schuler, W., Fosler-Lussier, E., Rosenfeld, A., & Danforth, D. (2015). Interpreting questions with a log-linear ranking model in a virtual patient dialogue system. *The Twelfth Workshop on Innovative Use of NLP foBuilding Educational Applications* (pp. 86–96). Stroudsburg, PA: The Association for Computational Linguistics.

Johnson, M., Lapkin, S., Long, V., Sanchez, P., Suominen, H., Basilakis, J., & Dawson, L. (2014). A systematic review of speech recognition technology in health care. *BMC Medical Informatics and Decision Making, 14*(1), 94.

Kelly, S. D. (2001). Broadening the units of analysis in communication: Speech and nonverbal behaviours in pragmatic comprehension. *Journal of Child Language, 28*(2), 325–349.

Khanna, A., Pandey, B., Vashishta, K., Kalia, K., Pradeepkumar, B., & Das, T. (2015). A study of today's AI through chatbots and rediscovery of machine intelligence. *International Journal of u-and e-Service, Science and Technology, 8*(7), 277–284.

Kisner, J. (2018). The technology giving voice to the voiceless. *The Guardian*. Available online https://www.theguardian.com/news/2018/jan/23/voice-replacement-technology-adaptive-alternative-communication-vocalid.

Koolagudi, S. G., & Rao, K. S. (2012). Emotion recognition from speech: A review. *International Journal of Speech Technology, 15*(2), 99–117.

Lala, R., Jeuring, J. T., & Overbeek, T. (2017). Analysing and adapting communication scenarios in virtual learning environments for one-to-one communication skills training. [ILRN2017] Available online http://castor.tugraz.at/doku/iLRN2017/iLRN2017paper34.pdf.

Lapakko, D. (1997). Three cheers for language: A closer examination of a widely cited study of nonverbal communication. *Communication Education, 46*(1), 63–67.

McMillan, D., Loriette, A., & Brown, B. (2015). Repurposing conversation: Experiments with the continuous speech stream. In *Proceedings of the 33rd annual ACM Conference on Human Factors in Computing Systems* (pp. 3953–3962). New York: ACM.

Newitz, A. (2015). *The Fembots of Ashley Madison*. Gizmodo. Available online: http://gizmodo.com/the-fembots-of-ashley-madison-1726670394.

Nolan, B. (2014). Extending a lexicalist functional grammar through speech acts, constructions and conversational software agents. In B. Nolan & C. Periñán-Pascual (Eds.), *Language Processing and Grammars: The Role of Functionally Oriented Computational Models* (pp. 143–164) Amsterdam, the Netherlands: John Benjamins Publishing Company.

Oh, A. H., & Rudnicky, A. I. (2000). Stochastic language generation for spoken dialogue systems. In *Proceedings of the 2000 ANLP/NAACL Workshop on Conversational Systems-Volume 3* (pp. 27–32). Morristown, NJ: Association for Computational Linguistics.

Ramos-Soto, A., Bugarin, A. J., Barro, S., & Taboada, J. (2015). Linguistic descriptions for automatic generation of textual short-term weather forecasts on real prediction data. *IEEE Transactions on Fuzzy Systems, 23*(1), 44–57.

Savin-Baden, M., Tombs, G., Burden, D., & Wood, C. (2013). "'It's almost like talking to a person': Student disclosure to pedagogical agents in sensitive settings. *International Journal of Mobile and Blended Learning, 5*(2), 78–93.

Reiter, E., & Dale, R. (1997). Building applied natural language generation systems. *Natural Language Engineering, 3*(1), 57–87.

Sidnell, J. (2011). *Conversation Analysis: An Introduction*. Hoboken, NJ: John Wiley & Sons.

Taylor, P. (2009). *Text-to-Speech Synthesis.* Cambridge, MA: Cambridge University Press.

Taylor, P., & Isard, A. (1997). SSML: A speech synthesis markup language. *Speech Communication, 21*(1), 123–133.

The Guardian. (2016). Microsoft 'deeply sorry' for racist and sexist tweets by AI chatbot. Available online https://www.theguardian.com/technology/2016/mar/26/microsoft-deeply-sorry-for-offensive-tweets-by-ai-chatbot

Thimm, M., Villata, S., Cerutti, F., Oren, N., Strass, H., & Vallati, M. (2016). Summary report of the first international competition on computational models of argumentation. *AI magazine, 37*(1), 102.

Vinyals, O., & Le, Q. (2015). A neural conversational model. *Proceedings of the International Conference on Machine Learning, Deep Learning Workshop.* Available online https://arxiv.org/abs/1506.05869

Wallace, R. (2003). *The Elements of AIML Style.* Alice AI Foundation. Available online https://files.ifi.uzh.ch/cl/hess/classes/seminare/chatbots/style.pdf.

Warwick, K., & Shah, H. (2016). Can machines think? A report on turing test experiments at the royal society. *Journal of experimental & Theoretical artificial Intelligence, 28*(6), 989–1007.

Wei, B., & Prakken, H. (2017). Defining the structure of arguments with AI models of argumentation. College Publications 68 1–22 Available online https://dspace.library.uu.nl/bitstream/handle/1874/356057/WeibinPrakken17.pdf?sequence=1.

Wen, T. H., Gasic, M., Mrksic, N., Su, P. H., Vandyke, D., & Young, S. (2015). Semantically conditioned lstm-based natural language generation for spoken dialogue systems. In *Proceeding of EMNLP* (pp. 1711–1721), Lisbon, Portugal, September.

Yamagishi, J., Veaux, C., King, S., & Renals, S. (2012). Speech synthesis technologies for individuals with vocal disabilities: Voice banking and reconstruction. *Acoustical Science and Technology,* 33(1), 1–5.

Zen, H., & Senior, A. (2014). Deep mixture density networks for acoustic modeling in statistical parametric speech synthesis. In *Acoustics, Speech and Signal Processing (ICASSP), 2014 IEEE International Conference* on (pp. 3844–3848). IEEE.

Zheng, T. F., Jin, Q., Li, L., Wang, J., & Bie, F. (2014). An overview of robustness related issues in speaker recognition. In *Asia-Pacific Signal and Information Processing Association, 2014 Annual Summit and Conference (APSIPA)* (pp. 1–10). IEEE.

CHAPTER 6

Architecture

INTRODUCTION

In the introduction to Part II, a component model of a virtual human was provided in order to provide a roadmap to help guide the reader through the different potential elements of a virtual human described in Chapters 3 through 5. This chapter will review some of the leading functional and cognitive architectures for creating a virtual human. The range considered will show the breadth of approaches, from those which are theoretical, through those that take an engineering approach, to those inspired by neuroscience. These models have been developed by cognitive computer scientists and others in order to both better understand human cognitive behaviour and functionality and to implement it effectively within computer-based systems.

BACKGROUND

The history of research into cognitive architectures dates to the 1950s and 1960s with Newell and Simon's relatively specialist Logic Theory Machine (1956) and Ernst and Newell's broader General Problem Solver (1967). Architectures became more sophisticated and even broader in scope with developments such as Adaptive Character of Thought (ACR) in the 1970s and State, Operator and Result (SOAR) in the 1980s. Laird (1992) presents a useful timeline of key cognitive architectures from 1975 to 2008 and identifies 3 main reasons for the creation of cognitive architectures:

- To explain the perceived behaviours and performance of the human, biological brain;

- To support the development of specialized AI agents that can interact with a particular environment; and
- To develop human-like agents with broad capabilities, i.e. virtual humans.

A particular architecture may be developed, or find itself being used to support, one, two or all of these goals.

Hilgard (1980) classifies mental activity into cognition (understanding), affection (emotion) and conation (purposefulness). Any virtual human model will probably need to ensure that all three of these are addressed – either as specific, discrete elements or regions of the models themselves, or as emergent behaviours/characteristics of the lower level functions within the architectures. Whilst researchers can talk about models forever, there is no substitute for actually building something. As Newell, one of the founding fathers of cognitive architectures, wrote 'any other way seems to involve just talk and exhortation, guaranteed to have little effect' (Newell, 1992).

This chapter will examine 10 models, representing a range of approaches in terms of how the cognitive functioning of a virtual human could be modelled.

- SOAR (Newell, 1990).
- Adaptive Character of Thought-Rational (ACT-R) (Anderson, 1996).
- Sloman's Human Cognition and Affect (H-CogAff) model (Sloman, 2001 and 2003) Cognitive Processing is published by Springer Berlin Heidelberg.
- Rodriguez's et al. (2010) neuroscience inspired architecture.
- Lin's (2011) EmoCOG architecture.
- OpenCog – an open source Artificial General Intelligence (AGI) initiative (Goertzel, 2008).
- The FearNot! Affective Mind Architecture (FAtiMA) and PSI models (Lim et al., 2012).
- Becker-Asano's (2014) WASABI model.
- The University of Southern California's Virtual Human toolkit (Gratch, 2013).
- The OpenAI Initiative (Brockman and Sutskever, 2016).

At the end of the chapter, the relative strengths and weaknesses of the models will be considered, particularly with regards to their potential for use in a practical virtual human implementation.

THE SOAR MODEL

SOAR (https://soar.eecs.umich.edu/), developed in the early 80s by John Laird, Allen Newell, and Paul Rosenbloom at Carnegie Mellon University, was an early attempt to develop a practical cognitive architecture which could be implemented as software to provide behaviours that mirrored those of human cognition.

In its original implementation, SOAR (Lehman et al., 1996) was a very general purpose problem-solving system. SOAR is goal-orientated, everything is about achieving one or more goals. Central to the system is the idea of 'goal context', which, for each problem to be solved, contains four elements.

- A goal – for example, in a game of cricket to get the batter out.
- The current state, which includes the various parameters which are important for solving the problem, for example, who is the batter, which is the current ball and over.
- The problem space, within which are defined or created all the possible states, each with their set parameter values, for example, the cricket pitch and all the players.
- The operators (i.e., actions) to take the agent from the current state to an adjoining new state in the problem space, for example, the type of ball to be bowled.

To solve a problem, SOAR obtains the relevant problem space and current state and places them in its working memory, and then goes through a process of 'elaboration' of all the rules (called production rules) in its long-term memory that define what operators could be 'suggested' and what parameters should be changed given the current state, and then enters a Decision phase to take the suggested operators (actions) and decide (based on a simple heuristic) which of the actions to take. If it does not have a clear course of action (referred to as an impasse), then it fires off a sub-goal to break the impasse, the results of which (typically a new production rule called a 'chunk') are then

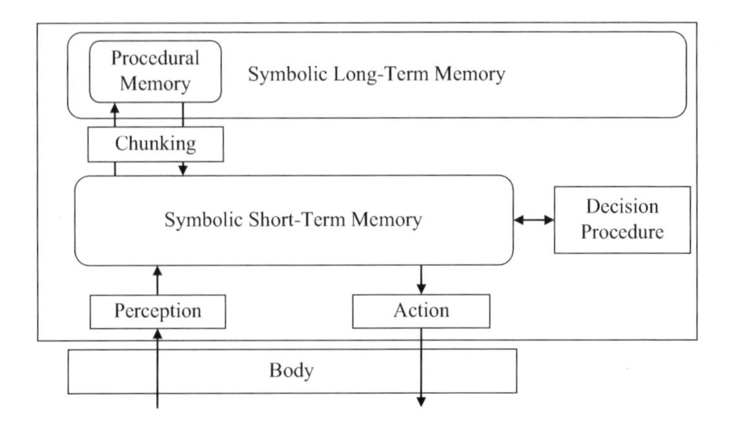

FIGURE 6.1 Original SOAR architecture. (After Laird (2008).)

added to its long-term memory so as to more easily deal with the situation should it arise again.

The overall architecture of this original version of SOAR is shown in Figure 6.1.

As can be seen, this is (deliberately) a very general purpose model and doesn't reflect any of the more specific virtual human functions.

However, as SOAR continued to be developed and applied to a variety of use-cases (and in particular to robotics), the architecture was extended to include many of these elements. SOAR 9 (the current version) includes extensions for:

- **Working Memory Activation** – effectively assigning meta-data to memories in terms of recentness and relevance – better mimicking human memory and to help select the most appropriate rules.
- **Reinforcement Learning** – so that operators which give good results are favoured next time around.
- **Emotion** – with an appraisal detector so that appraisals lead to emotions, emotions influence mood, and mood and emotion determine feelings (Marinier and Laird, 2007).
- **Semantic Memory** – to model a wide variety of human data so that SOAR agents can reason and use general knowledge about the world.

- **Episodic Memory** – to contain memories of what was experienced over time (Tulving, 1983).
- **Visual Imagery** – to better support vision and imagery-related problems.
- **Clustering** – to detect patterns in experiences and create new symbols that represent those patterns (Granger, 2006).

In its extended form, SOAR covers many of the elements identified in the Virtual Human Component Model, and the architecture is probably flexible enough to represent the rest. SOAR has been used to implement agents to play many computer games (such as StarCraft, Quake II, Descent 3, Unreal Tournament, and Minecraft) and has also been used to simulate face-to-face dialogues within a virtual environment trainer developed at the USC Institute of Creative Technology. However, it seems that much of the academic and development focus of cognitive architectures has now moved away from SOAR.

ACT-R

ACT-R (http://act-r.psy.cmu.edu/), originally developed by John Anderson at Carnegie Mellon University (Anderson, 1996) is a superficially similar Cognitive Architecture, similarly developed initially on the LISP (List Processor) programming language and at a similar time. It also features declarative knowledge, facts, and procedural knowledge rules of how to do various cognitive tasks – and, indeed, this separation is more of a core feature of ACT-R than it is of SOAR. A defining difference, though, is that SOAR appears to have been designed top-down – building a system to generate the right responses and being based on the 'general characteristics of intelligent agents, rather than detailed behavioural phenomena' (Johnson, 1997), whereas ACT-R appears to have a more bottom-up approach, growing 'out of detailed phenomena from memory, learning, and control' (Johnson, 1997).

The latest version of ACT-R, released in 2005 (ACT-R 6.0) includes modules to support: audio input and speech output (enabling conversation support), visual input, motor control output (enabling embodied robotic support), alongside the core modules of procedural memory, declarative memory, goal management and an 'imaginal' module – effectively a short-term memory. Recent developments of interest with ACT-R have included extending it to better support social simulation and communication (Smart et al., 2014) and integrating it with the Unity game engine (Smart, 2016). ACT-R has also been heavily used in support of the creation of cognitive tutors to support learners (Ritter et al., 2007).

SLOMAN'S H-COGAFF MODEL

Aaron Sloman of the University of Birmingham developed the CogAff and H-CogAff models (http://www.cs.bham.ac.uk/research/projects/cogaff/) during the 2000s. Unlike the SOAR and ACT-R models, his starting point was emotions rather than decision making, and the models were more of a philosophical tool than a blueprint architecture intended for coding.

Sloman's starting point was the 'triple-tower' model of perceptual and action mechanisms developed by Nilsson (1998) and consisting of:

- Perception mechanisms;
- A central processing function; and
- Action mechanisms.

Sloman then developed a three-layer evolutionary model of information processing, which from bottom to top described:

- The oldest, reactive processing mechanisms, for example, instinctive reactions;
- Newer, deliberative, 'what-if' mechanism, for example, simple problem-solving;
- The newest reflective, meta-management processes, for example, what did I learn?

Overlaying this layered model onto Nilsson's towers model produced the Cog-Aff (Cognitive Affective) model (Sloman, 2001) shown in Figure 6.2.

Sloman is at pains to point out that the model isn't a linear left-to-right pipeline, or top-to-bottom dominance, instead, any, or multiple, parts may be dominant at any one time, and exchanging information with any other part.

Sloman (2003, 19) saw Cog-Aff as:

> [N]ot an architecture, but a generic schema, something like a grammar which covers a variety of different cases. Just as a

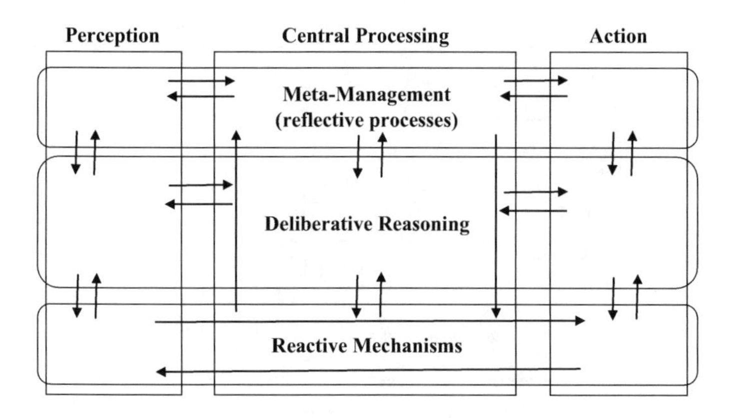

FIGURE 6.2 The basic Cog-Aff model. (After Sloman, A., *Cogn. Process.*, 2, 177–198, 2001.)

> grammar distinguishes types of sentence-components and specifies how they can be combined, so our architecture schema, distinguishes types of components of the architecture, for example, reactive, deliberative and meta-management, and ways in which they can be combined.

He also pointed out that not every component specified in it need be present in every implementation. So, by using just the reactive layer the cognitive functions of a simple insect could be modelled, or by using both reactive and deliberative layers a typical mammal could be modelled.

H Cog-Aff (Human Cog-Aff) is a particular instantiation of the generic Cog-Aff schema to model human cognition and is shown in Figure 6.3. As well as using all three towers and layers, this also adds an additional features such as long-term memory, motive activation and personae (controlling parameters to implement different cultural behavioural norms).

Coming back to Sloman's interest in emotions, the model is designed to support three different classes of emotion as follows (Sloman, 2001):

- **The reactive layer**, including the global alarm mechanism, accounts for primary emotions (for example, being startled, frozen with terror, sexually aroused);
- **The deliberative layer** supports secondary emotions like apprehension and relief which require 'what if' reasoning abilities (these are semantically rich emotions); and

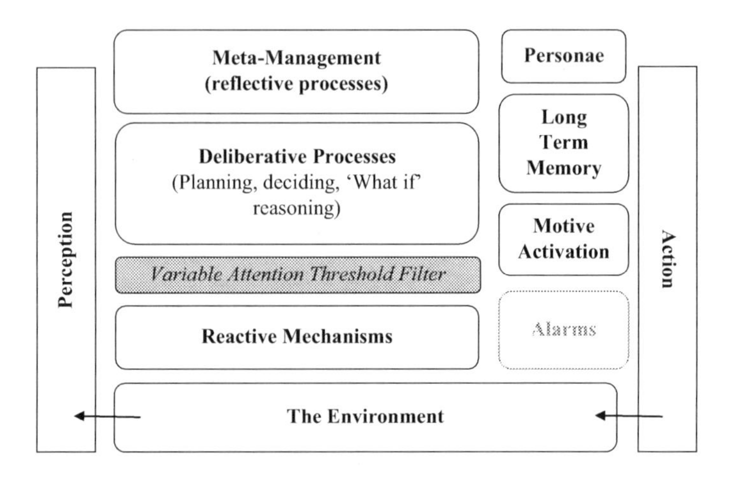

FIGURE 6.3 Sloman's H Cog-Aff architecture. (After Sloman (2009).)

- **The meta-management (reflective) layer** supports not only control of thought and attention but also the loss of such control, as found in typical human tertiary emotions, such as infatuation, humiliation, thrilled anticipation of a future event.

Whilst the core Cog-Aff/H Cog-Aff project ceased in 2004, the spirit of it has lived on in the 2004–2008 CoSy (http://www.cs.bham.ac.uk/research/projects/cosy/partners/) and 2008–2012 CogX (http://cogx.eu/) European Union (EU) funded research projects.

OPENCOG

OpenCOG is a software framework designed specifically for the integration of AI algorithms, and is intended to support the construction of integrative AI systems from component AI algorithms and structures (Hart and Goertzel, 2008). It was created in response to a feeling that (then) current AI research was too focussed on isolated capabilities, rather than trying to build a 'complete' AI. The original codebase grew out of work by Ben Goertzel's Novamente company, and the initial focus was on creating the AI 'brain' for humanoid robots created by Hanson Robotics.

Hart and Goertzel (2008) divide an AI system design into four components.

- **Cognitive Architecture:** the overall design of an AI system; what parts does it have, and how do they connect to each other.

- **Knowledge Representation:** how the AI system internally stores declarative, procedural and episodic knowledge, and how creates its own representations for elemental and abstracted knowledge in new domains it encounters.
- **Learning:** how the AI system learns new elemental and abstracted knowledge, and how it learns how to learn.
- **Teaching Methodologies:** how the AI system is coupled with other systems so as to enable it to gain new knowledge about itself, the world and others.

OpenCog addresses each of these through:

- An architecture-neutral, modular cognitive architecture.
- AtomTable, a graph-based memory system for knowledge representation and storage, supporting both connectionist and symbolic AI approaches;
- Learning through probabilistic program evolution and probabilistic logical inference; and
- A virtual world (AGISim - http://sourceforge.net/projects/agisim) in which to learn and teach the embodied AI, and a natural language system to support direct teaching.

Figure 6.4 provides an overview of the OpenCOG cognitive architecture.

Goertzel et al. (2014) describes OpenCOG as having the following key components:

- A Perception Synthesizer (PS) to gather sensory information from the robot;
- An Action Orchestrator (AO) to decide what action to take on the basis of the PS information;
- An Action Creator (AC) to enact the decided actions in the robot;
- An Action Data Source (ADS), which provides a library of actions for the AC;

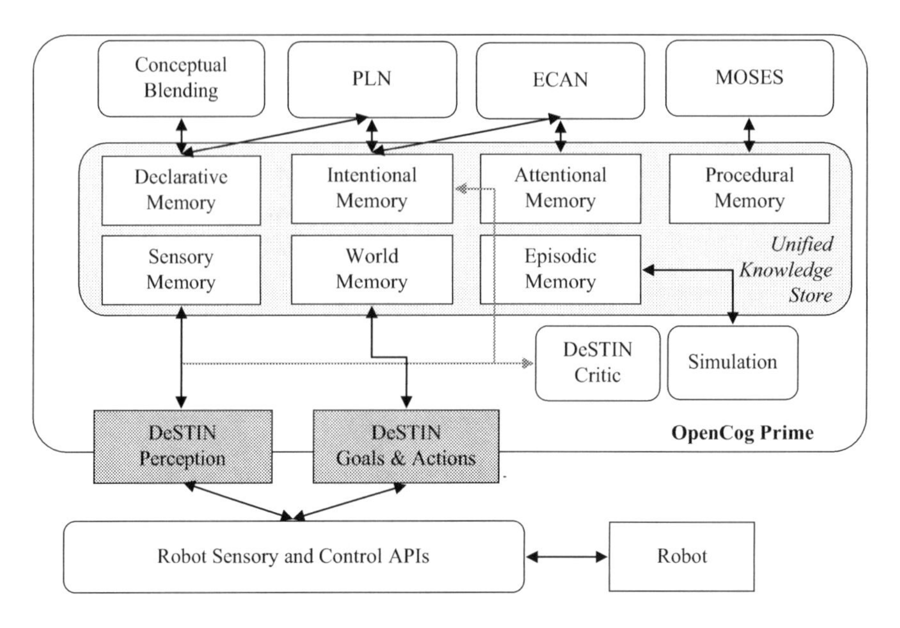

FIGURE 6.4 OpenCOG model. (After https://wiki.opencog.org/w/OpenCog Prime:Cognitive_Architecture.)

- A Robot Controller (RC), which provides low level, autonomous robot control, for example, gaze control; and
- A Robot Control User Interface (RCUI) for manual control.

OpenCOG's memory is based on an 'AtomTable' – a graph-based memory store. Goertzel and Duong (2009, 65) states that:

> One of the core principles underlying OpenCogPrime is 'cognitive synergy', which states that an intelligent system should have different cognitive processes corresponding to each type of memory, and that these processes should interact synergetically, so that when one of them gets stuck, it can appeal to the others for help.

In OpenCOG, six different memory types are identified:

- Declarative,
- Procedural,
- Episodic,

- Attentional – related to importance and forgetting/memory resource management,
- Intentional – related to goals and planning, and
- Sensory – related to sensory input.

OpenCOG originally laid out an interesting roadmap for their development (https://wiki.opencog.org/w/OpenCogPrime:Roadmap) comprised of six phases.

1. Basic Components.
2. Artificial Toddler.
3. Artificial Child.
4. Artificial Adult.
5. Artificial Scientist.
6. Artificial Intellect.

In 2015 (when the wiki page was last updated), they reported that they were 'near the start of Phase Two, and still wrapping up some aspects of Phase One'. Updates were still being made to the OpenCOG GitHub repository (https://github.com/opencog/opencog) in 2018, although, in 2017, there were only two news updates on the main website and only one blog post.

RODRIGUEZ'S NEUROSCIENCE MODEL

Rodriguez's neuroscience model takes a very different approach in that it seeks to develop a direct mapping between the functional areas of the brain and the functional areas of the model. Rodriguez argues that 'since each of our components works according to a specific neuroscience theory, the resulting behaviour must resemble those of humans' (Rodriguez, 2010). His model is shown in Figure 6.5, where the functional areas of the brain are mapped onto the functional areas of the system.

This approach does not appear to have been widely adopted elsewhere, possibly since the highly interconnected nature of the brain means that it is not possible to say that Area X is the only area driving Function Y, nor that Area X only drives Function Y – and indeed Figure 6.5 does show

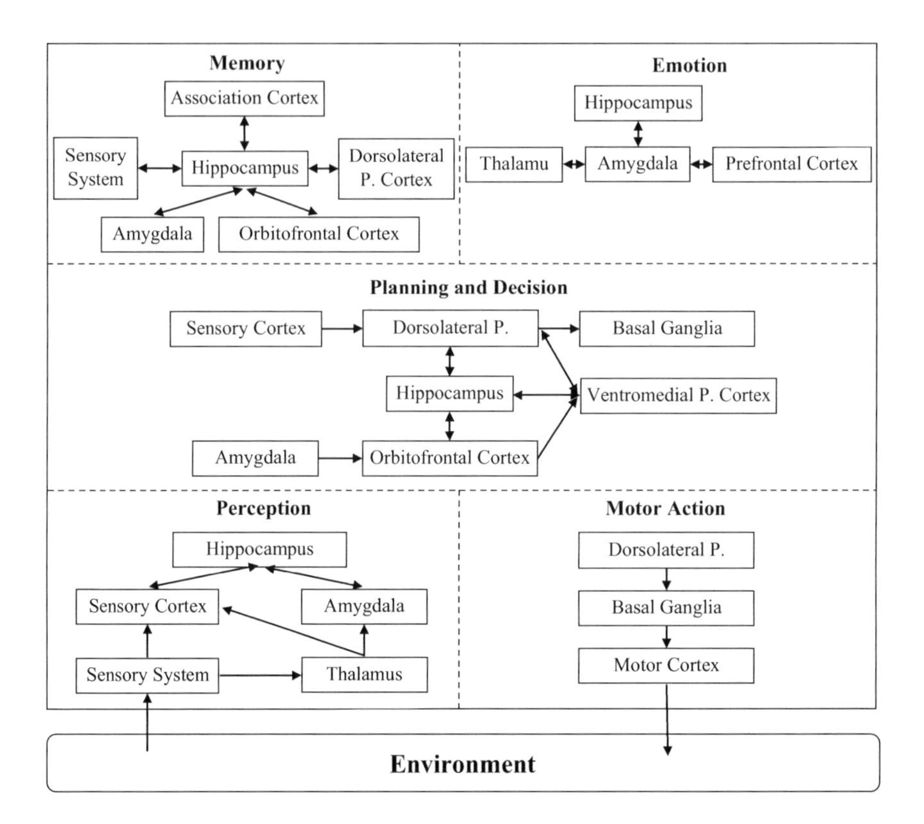

FIGURE 6.5 Rodriguez's neuroscience model. (After Rodriguez, L. F. et al., A cognitive architecture based on neuroscience for the control of virtual 3D human creatures, *International Conference on Brain Informatics*, 328–335, 2010.)

replication of areas across functions (for instance, the amygdala appears in four of the five functions. However, the model does provide a useful reminder of how the human brain achieves the ends which a virtual human is seeking to replicate and can help to inform other models.

LIN'S EMOCOG ARCHITECTURE

Whereas cognitive architectures such as SOAR and ACT-R appear to have had emotional elements bolted on at a later date, Lin's EmoCOG architecture (Lin, 2011) has emotion as a key driver. Building on the work of Bechara and Damasio (Bechara et al., 2000), emotions are seen as not only a necessity for decision-making but also closely linked to long-term memory through the levels of emotional arousal. Specifically, EmoCOG is based on appraisal theory which 'generally argues that people are constantly evaluating their environment, and that evaluations result in

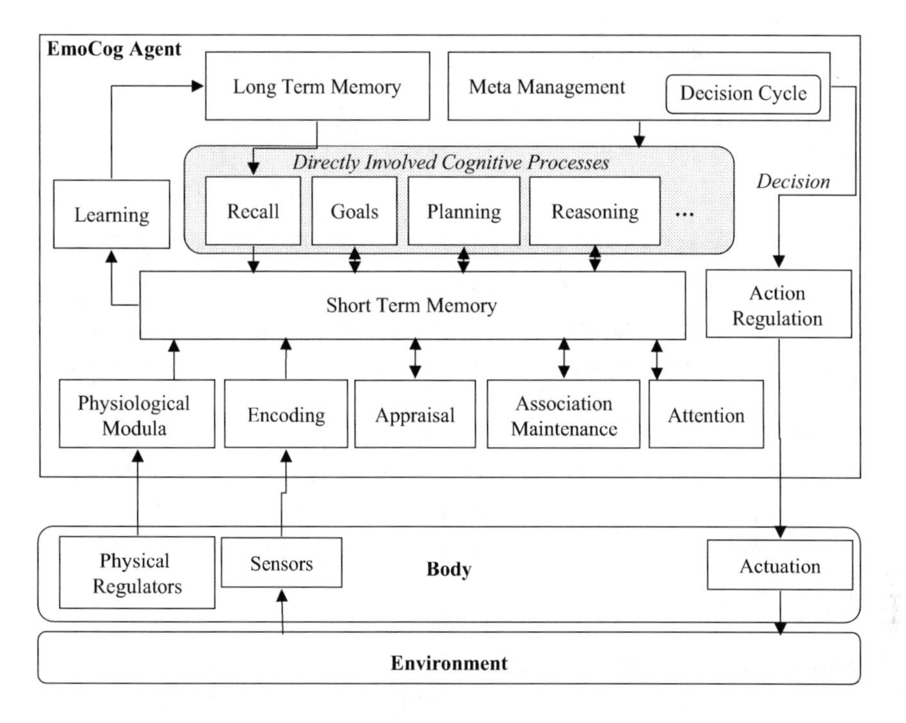

FIGURE 6.6 EmoCOG model. (After Lin, J. et al., EmoCog: Computational integration of emotion and cognitive architecture, In *Proceedings of the Twenty-Fourth FLAIRS Conference*, 2011.)

emotions such as fear or anger' (Lin, 2011), and which has been defined in appraisal theory models such as OCC (Ortony et al., 1990). The EmoCOG model is shown in Figure 6.6.

Of particular note is that EmoCog's short-term memory is based on a network of memory nodes and association links. Based on the ideas of Bower (1981) and Anderson (1983), these associations have a strength which can be reinforced (by use) or decay (by neglect). In addition, each node carries values for current arousal, remembered arousal (an average over time), current valence (the degree of like/dislike) and remembered valence (average over time), recentness of recall and how many times the node has been accessed.

When a new event is perceived, the attention module selects the memory nodes that will be retrieved, by first finding the node with the greatest current arousal, utility and urgency, and then also retrieving the nodes directly and indirectly (via the weighted associations) associated with this first node out to a certain threshold. In this way, the EmoCog model simulates the way that on seeing a particular sight, sound or smell we may have a whole cascade of seemingly unconnected memories being triggered.

THE FEARNOT! AFFECTIVE MIND ARCHITECTURE AND PSI MODELS

Lim's FearNot! Affective Mind Architecture (FAtiMA) (Lim et al., 2012) was created in order to 'create believable autonomous Non-Player Characters (NPCs) in video games in general and educational role play games in particular' (Lim et al., 2012). The resultant NPC integrates two separate models: the cognitive appraisal-based FAtiMA architecture (Dias and Paiva, 2005), and the drives-based PSI model. Whereas traditionally, game NPCs have been driven by finite state machine – they move between different states such as guard-patrol-chase-hide, Lim's work sought to create more realistic NPCs by giving them goals to achieve, actions to achieve the goals, and the ability to respond realistically to their environment. FAtiMA includes a deliberative component based on the Belief-Desire-Intent (BDI) model and a reactive component based on the Ortony, Clore, & Collins (OCC) appraisal model, both described in Chapter 4. The FAtiMA model is shown in Figure 6.7.

Lim et al. (2012, 295) notes that:

> FAtiMA does not take into account the physiological aspects of emotion. Furthermore, since it follows the BDI model closely and does not extend it with additional concepts that justify the appearance of goals/desires, FAtiMA faces the known problems of the BDI model.

BDI issues include no grounding as to what an agent's desires should be (or why), no reason why one desire should be preferred over another, and a tendency not to learn from experience or adapt to their environment. To address these issues, Lim describes how the PSI model (Dorner, 2003) is used alongside FAtiMA. The PSI model is shown in Figure 6.8.

The PSI model introduces the concept of 'needs' to the virtual human, and it can be seen from Figure 6.8 that they bear some semblance to Maslow's well-known hierarchy of needs discussed in Chapter 4. PSI also introduces the idea of emotional arousal and resolution (deliberative/accuracy) levels which help to further define the personality of the virtual human.

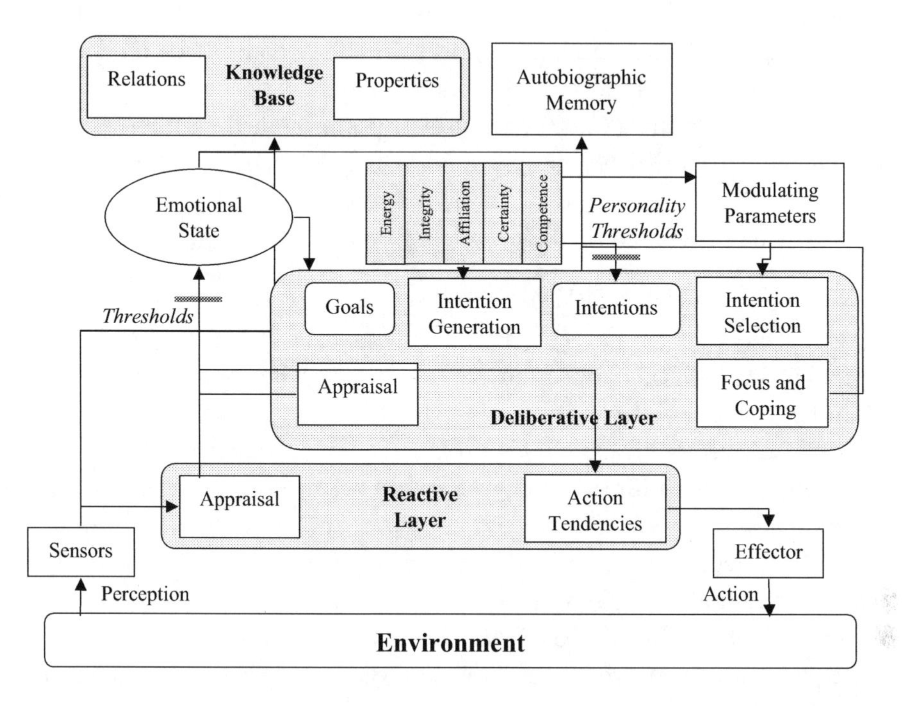

FIGURE 6.7 FAtiMA/PSI model. (After Lim, M.Y. et al., *Auton. Agents Multi. Agent Syst.*, 24, 287–311, 2012.)

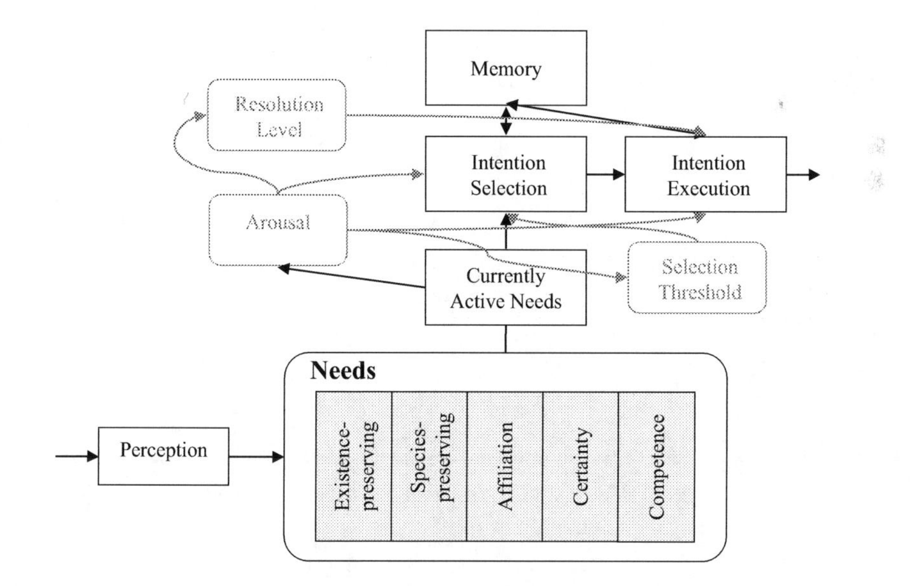

FIGURE 6.8 The PSI model. (After Dorner, D. The mathematics of emotions. In F. Detje & H. Schaub, Eds., *Proceedings of the Fifth International Conference on Cognitive Modeling*, 75–79, Bamberg, Germany, Universitäts-Verlag Bamberg, 2003.)

Martínez-Miranda et al. (2004) describes how the FAtiMA model was used in the Help4Mood virtual agent to support people with depression, and Gomes and Jhala (2013) describes its use in modelling interpersonal conflict. FAtiMa also provides the Emotion Engine in the Triple Embodied Conversational Agent model (Kiryazov and Grinberg, 2010), and in Ho's (2008) narrative based Intelligent Virtual Agents.

BECKER-ASANO'S WASABI MODEL

Having moved from decision driven systems to emotion-driven systems, the WASABI Affect Simulation Architecture for Believable Interactivity (Becker-Asano, 2014) goes the next step by considering the importance of the link between emotion and mood. A pleasant event not only results in a happy emotion but also a lift in mood, which then predisposes the agent towards future pleasant events. But, if no further pleasant events occur, the agent's mood gradually drifts back to a neutral state. The 'space' within which the movement of emotion and mood takes place is referred to as the Pleasure Arousal Dominance, or PAD, space (Russell and Mehrabian, 1997). In this space, valence (pleasure/displeasure) is on the X-axis, low/high arousal on the Y-axis, and submissive/dominant on the Z axis – as shown in Figure 6.9.

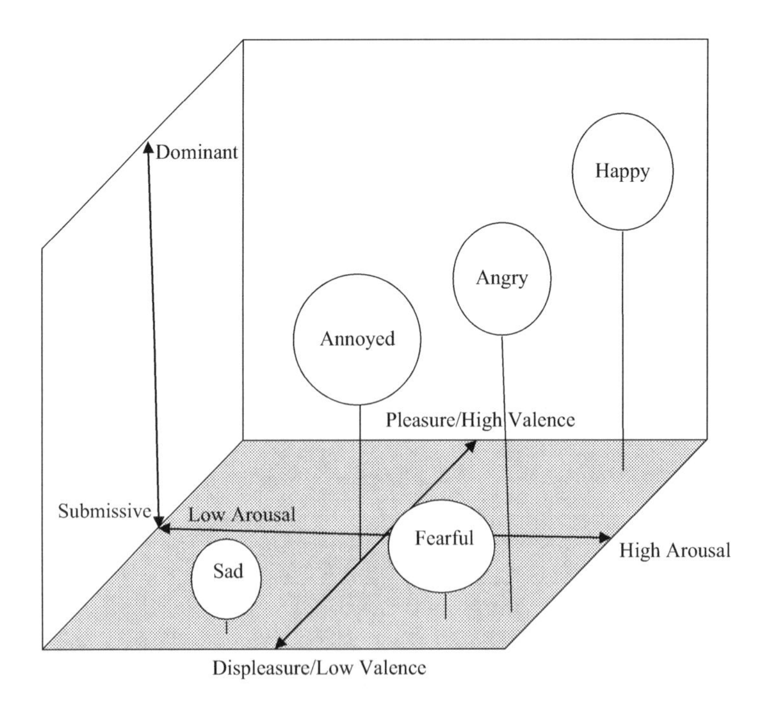

FIGURE 6.9 The pleasure arousal dominance (PAD) space.

In the WASABI implementation of the PAD space, each emotion and mood is represented by a region of the PAD space, some of which overlap, and the current coordinates of the agent in the PAD space define its current mood/emotion. As well as the primary emotions of happy, sad, fear, annoyed/disgust and angry, and the 'extra' emotion of surprise, WASABI also implements the secondary emotions of hope, relief and 'fears-confirmed'.

Figure 6.10 shows the core WASABI architecture. It includes the concept of conscious and unconscious appraisal and deliberative and involuntary actions as found in Sloman's H-CogAff Architecture, as well as the Pleasure Arousal Dominance space emotion/mood elements.

Figure 6.11 shows how WASABI could be integrated into a typical virtual human architecture, and it can be seen that it is designed to operate alongside a cognition/decision module – WASABI providing the emotion and mood modelling.

WASABI has been used in a range of virtual human applications, including the MAX card playing chatbot (Becker et al., 2004), the MARC character animation system (Courgeon et al., 2008) and the Scitos-G5 robot (Klug and Zell, 2013).

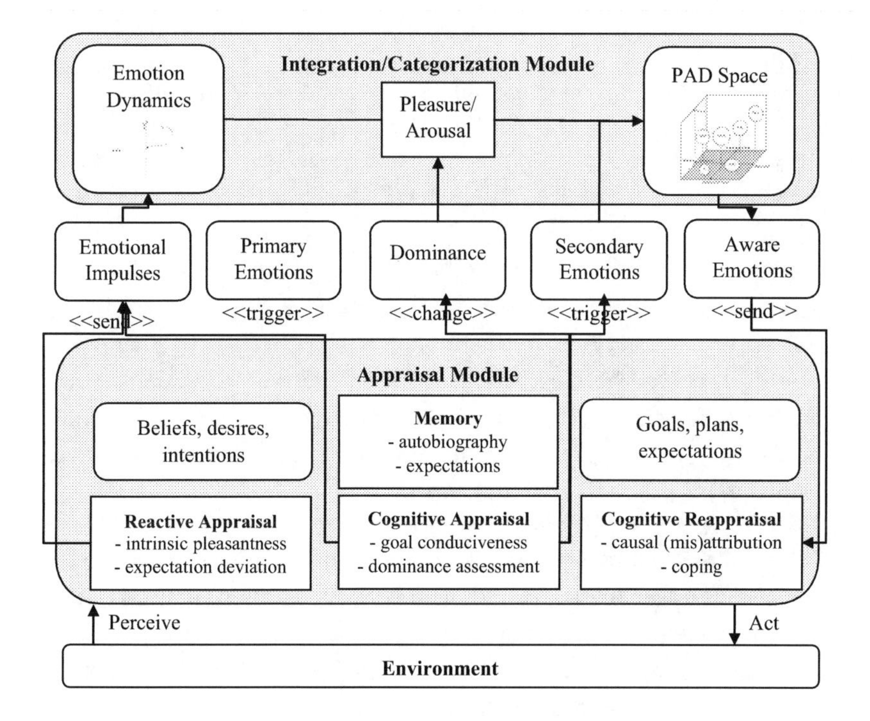

FIGURE 6.10 The WASABI architecture.

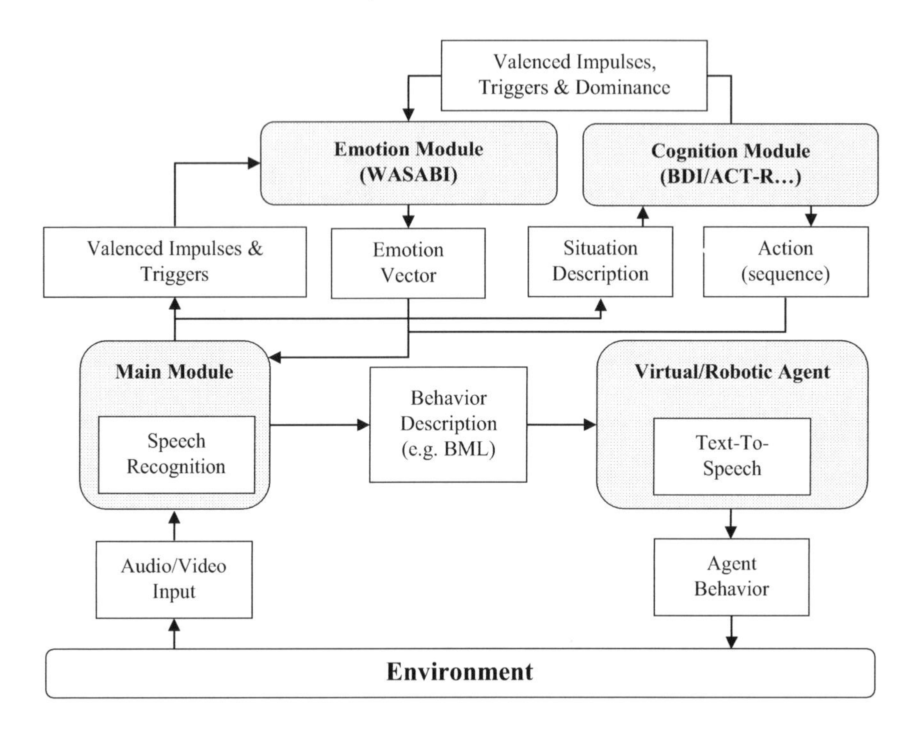

FIGURE 6.11 WASABI integrated into a virtual human. (After Becker-Asano, C., WASABI for affect simulation in human-computer interaction, In *Proceeding International Workshop on Emotion Representations and Modelling for HCI Systems*, 2014.)

UNIVERSITY OF SOUTHERN CALIFORNIA'S VIRTUAL HUMAN TOOLKIT

The Virtual Human Toolkit is, as its name suggests, a toolkit for virtual human research made available by University of Southern California's (USC) Institute for Creative Technologies (ICT) (Gratch et al., 2002). Although informed by cognitive science, this is not an academic model or architecture of human cognition, but rather a set of software elements which work together to create the system needed to control a virtual human (VH). As Gratch et al. (2013, 2) describes it:

> [T]he Virtual Human Toolkit is a general-purpose collection of integrated VH capabilities, including speech recognition, natural language processing, perception, and nonverbal behavior generation & execution. The goal of the Virtual Human Toolkit is to make creating VHs easier and more accessible and, thus expand the realm of VH research as well as other research areas, including cognitive science.

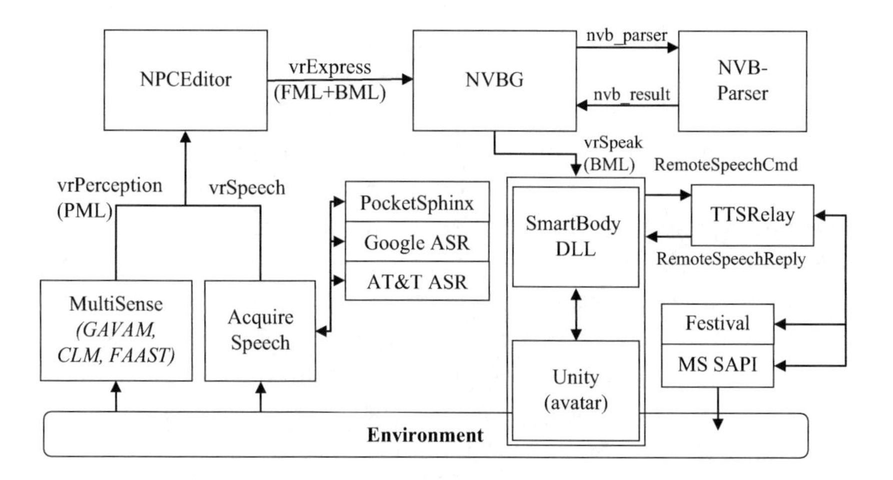

FIGURE 6.12 USC virtual human toolkit. (After Hartholt, A.D. et al., All together. now, In *Proceedings of the 13th International Conference on Intelligent Virtual Agents*, pp. 368–381, Edinburgh, UK, August 29–31, Springer, Berlin, Germany, 2013.)

The system was developed out of a cross-disciplinary workshop held in 2002, with the aim of developing a modular architecture and interface standards that would allow researchers in this area to collaborate and share each other's work.

The architecture of the USC Virtual Human Toolkit is shown in Figure 6.12.

In comparison to the other architectures so far described, this architecture is heavily implementation focused, with software systems or interfaces identified for each element (Hartholt et al., 2013).

- Markup language-based interface definitions for Perception (PerceptionML) and Expression (FunctionalML and BehaviouralML).
- The NPCEditor providing the natural language system through a statistical text classification algorithm.
- The virtual human body is instantiated in the Unity game engine.
- The Non-Verbal Behaviour Generator generates non-verbal behaviour (!) based on the virtual human's chosen natural language responses (for example, shake of head if saying 'no').
- Speech recognition is provided through either Google, AT&T or Pocket Phoenix Speech Recognition systems.

- The open source Festival and Microsoft Speech Application Programming Interface (MSAPI) are used to provide text to speech.

The design and composition of the Toolkit reflect the fact that its main use is in the creation of virtual humans for training systems (Swartout 2006 and Hartholt et al., 2013). The architecture, as it stands, does not include the elements for 'higher-level' functions, such as memory, emotion, reasoning and decision-making seen in most of the other architectures, although these can, of course, be added. In fact, the same team has described how OCC type-appraisal/emotion models can be implemented (Gratch et al., 2002) using the Behaviour Expression Animation Toolkit (BEAT), and Swartout (2006) describes the use of SOAR for a predecessor to the VH Toolkit within the context of virtual humans for a Mission Rehearsal Exercise simulation for the United States Department of Defense.

OPENAI

One of the more recent initiatives in AI development is OpenAI – whose founders include Elon Musk. OpenAI declares itself to be:

> [A] non-profit artificial intelligence research company. Our goal is to advance digital intelligence in the way that is most likely to benefit humanity as a whole, unconstrained by a need to generate a financial return. Since our research is free from financial obligations, we can better focus on a positive human impact. (Brockman and Sutskever, 2015)

OpenAI aims to build safe Artificial General Intelligences and ensure that the benefits of AI are distributed widely across society through being an open-source endeavour. Artificial General Intelligence is explored in more detail in Chapter 13. The founders of OpenAI have pledged $1bn to finance the company, although they 'expect to only spend a tiny fraction of this in the next few years' (Brockman and Sutskever, 2015).

Focusing primarily on deep learning techniques to create an Artificial General Intelligence, OpenAI identifies three initial goals or projects to focus its work:

1. Building a better brain for a useful and efficient house robot – to drive a commercial physical robot.

2. Building an agent with useful natural language understanding – although this is couched in terms of performing 'a complex task specified by language and ask for clarification about the task if it's ambiguous', with other challenges such as carrying out a conversation or being able to fully understand a document also being identified.

3. Solving a wide variety of games using a single agent (Sutskever, 2016).

OpenAI has published a large number of papers on its work and is making most of the code available through services such as GitHub. This includes the code for its Gym toolkit (Brockman et al., 2016) for testing and comparing reinforcement learning algorithms, and its Gym-based Universe platform for measuring and training an AI's general intelligence by enabling it to be interfaced to almost any game, website and other application. OpenAI appears to be taking a 'weak AI' approach. There is no overarching architecture as with the other models presented in this chapter, but through solving a variety of problems and addressing each of the goals detailed above – largely through the use of deep learning techniques – the intent appears to be to evolve a generic problem-solving approach, code base or AI which could ultimately display generalized intelligence.

INTEGRATION STANDARDS

Whilst not a model *per se*, it is important to note that attempts have been made to develop standards for virtual humans that would make it easier to 'mix and match' different technologies in order to create a 'best of breed' solution. Whilst there are some dangers in this, since many sub-systems operate better when closely coupled, it does provide a more pragmatic approach to virtual human development.

Whilst there does not appear to be a single, or even small group of, standards in common *de facto* or *de jure* use, there are a number which could be of potential use to virtual human development and which are summarised in Table 6.1.

There ought to be a role for interconnection and, particularly, interface standards within virtual human development, but at the present time, this does not seem to be a priority with the community.

CRITIQUE AND FUTURES

No one model appears to provide a 'complete' architecture for a virtual human, with each focusing on one or two particular areas of functionality at the cost of others. This is perhaps in keeping with many of them

TABLE 6.1 Virtual Human Interface Standards

Interface Standard	Description
Virtual Human Markup Language (VHML) (Marriott et al., 2001)	Built on existing standards, such as those specified by the W3C Voice Browser Activity, and added new tags to accommodate functionality for Virtual Humans. XML/XSL based and included sub-standards to cover areas such as Dialogue Management, Facial and Body animation, speech, emotion and gesture. Whilst broad and laudable in its goals, VHML does not appear to have seen any activity for a decade.
Discourse Plan Markup Language (DPML) and Affective Presentation Markup Language (APML) (De Carolis et al., 2004).	Discourse Plan Markup Language (DPML) and Affective Presentation Markup Language (APML) work together to take the description of what an avatar wants to do (DPML) and enrich it to provide a set of action instructions (APML) for the avatar's body.
Function Markup Language (FML) and Behaviour Markup Language (BML) Kopp et al. (2006)	Part of the SAIBA architecture where FML describes intent without referring to physical behaviour, and BML describes the desired physical realization of the actions.
Perception Markup Language (PML) (Scherer et al., 2012).	PML is inspired by FML and BML and attempts a standardized representation of perceived nonverbal behaviours. The work was funded in part by DARPA.
Character Markup Language (CML) (Arafa and Mamdani, 2003)	Character Markup Language (CML) is high level standard for body and facial control and leverages the MPEG-4 standard for avatar control.
Avatar Markup Language (AML) (Kshirsagar et al., 2002)	The Avatar Markup Language (AML) encapsulated Text to Speech, Facial Animation and Body Animation in a unified manner with appropriate synchronization, and as with CML could be used to drive MPEG-4 animations.
Avatar Sensing Markup Language (ASML) and Avatar Action Markup Language (AAML) (Slater and Burden, 2009)	ASML provides a way of defining the sensory events (text-speak heard, object appearance/approach, weather etc.) and passing them from the body (avatar) element to the brain element. Its counterpart, AAML, provides a way of defining actions (and sequences of actions) to be passed from a virtual human brain to a manifestation of the virtual human (for example, avatar in a 3D world, chatbot on a 2D web page).

(particularly, the early ones) being more investigative models of cognition or explorations of specific tasks, such as problem-solving, with only the more recent models adopting a more holistic approach.

In terms of being a model that an engineer could actually work with to create a virtual human, then the USC Virtual Human Toolkit and the fuller WASABI would appear to come the closest. Both, though, are missing a detailed memory model, which could be provided by OpenCOG, and the PSI model could provide a structure for longer-term motivation that is absent from most of the others.

It is evident that 'cognition' itself does not appear in any model as they are typical cognitive architectures, with cognition itself being typically defined as 'all of the mental activities involved in receiving information, comprehending it, storing it, retrieving it, and using it' (Ylvisaker et al., 2006), and so represents the sum of the activities going on. Cognition (and possibly reasoning) may be an emergent feature of the virtual human, and if the virtual human is actually aware of its cognition and reasoning, then it could be seen to be on the route to self-awareness and sentience. This will be investigated in more detail in Chapter 13.

CONCLUSION

This chapter has examined a variety of popular and interesting cognitive architectures. Each of them attempts to explore how the mind, human or computer, could be constructed at a functional level in order to generate the behaviours that would be expected to be seen in a physical or virtual human. The chapter has explored how different architectures have had different areas of focus, from more abstract approaches in the earlier models, such as ACT-R and SOAR, through biologically inspired models, such as Rodriguez's, to very practical implementations, such as USC Virtual Human Toolkit.

The models have, by their very nature, focused on a relatively disembodied mind. Whilst the diagrams may show inputs and outputs from and to the external world, the place of the cognitive architecture, the mind, in some wider world, is not a direct concern. However, there is a strong body of thought that sees it as essential to consider the mind in relation to its environment, an embodied mind, if intelligence, and especially a human-like intelligence, is to be created. Chapter 7 explores the arguments around embodiment in more detail.

REFERENCES

Anderson, J. R. (1983). A spreading activation theory of memory. *Journal of Verbal Learning and Verbal Behavior, 22*(3), 261–295.

Anderson, J. R. (1996). ACT: A simple theory of complex cognition. *American Psychologist, 51*(4), 355.

Arafa, Y., & Mamdani, A. (2003). Scripting embodied agents behaviour with CML: character markup language. In *Proceedings of the 8th International Conference on Intelligent User Interfaces* (pp. 313–316). New York: ACM.

Bechara, A., Damasio, H., & Damasio, A. R. (2000). Emotion, decision making and the orbitofrontal cortex. *Cerebral Cortex, 10*(3), 295–307.

Becker, C., Kopp, S., & Wachsmuth, I. (2004). Simulating the emotion dynamics of a multimodal conversational agent. In A. E, Dybkjær., L, Minker., W, Heisterkamp. (Eds) *Proceedings of the Tutorial and Research Workshop,* ADS 2004, Lecture Notes in Computer Science (LNAI) (Vol. 3068, pp. 154–165). Berlin, Germany: Springer.

Becker-Asano, C. (2014). WASABI for affect simulation in human-computer interaction. In *Proceeding International Workshop on Emotion Representations and Modelling for HCI Systems.* New York: ACM.

Becker-Asano, C., & Wachsmuth, I. (2010). Affective computing with primary and secondary emotions in a virtual human. *Autonomous Agents and Multi-Agent Systems, 20*(1), 32.

Bower, G. H. (1981). Mood and memory. *American Psychologist, 36*(2), 129.

Brockman, G., Cheung, V., Pettersson, L., Schneider, J., Schulman, J., Tang, J., & Zaremba, W. (2016). OpenAI gym. Available online arXiv preprint arXiv:1606.01540.

Brockman, G., & Sutskever, I. (2015). Introducing OpenAI [Web log post]. Available online https://blog.openai.com/introducing-openai/

Courgeon, M., Martin, J. C., & Jacquemin, C. (2008). Marc: A multimodal affective and reactive character. In *Proceedings of the 1st Workshop on Affective Interaction in Natural Environments* (pp. 12–16). New York: ACM.

De Carolis B., Pelachaud C., Poggi I. & Steedman M. (2004) APML, a markup language for believable behavior generation. In H. Prendinger and M. Ishizuka (Eds) *Life-Like Characters. Cognitive Technologies* (pp. 65–85). Berlin, Germany: Springer.

Dias J. & Paiva A. (2005). Feeling and reasoning: A computational model for emotional characters. In C. Bento, A. Cardoso & G. Dias (Eds.), *Progress in Artificial Intelligence. EPIA 2005. Lecture Notes in Computer Science* (Vol. 3808, pp. 127–140). Berlin, Germany: Springer.

Dorner, D. (2003). The mathematics of emotions. In F. Detje & H. Schaub. (Eds.), *Proceedings of the Fifth International Conference on Cognitive Modeling* (pp. 75–79). Bamberg, Germany: Universitäts-Verlag Bamberg, April 10–12.

Ernst, G. W., & Newell, A. (1967). Some issues of representation in a general problem solver. In *Proceedings of the April 18–20, 1967, Spring Joint Computer Conference* (pp. 583–600). New York: ACM.

Goertzel, B., & Duong, D. (2009). OpenCog NS: A deeply-interactive hybrid neural-symbolic cognitive architecture designed for global/local memory synergy. In *AAAI Fall Symposium: Biologically Inspired Cognitive Architectures* (pp. 63–68). Palo Alto, CA: AAAI.

Goertzel, B., Hanson, D., & Yu, G. (2014). Toward a robust software architecture for generally intelligent humanoid robotics. *Proceeding Computer Science, 41*, 158–163

Gomes, P. F., & Jhala, A. (2013). AI authoring for virtual characters in conflict. In *Proceedings on the Ninth Annual AAAI Conference on Artificial Intelligence and Interactive Digital Entertainment.* Palo Alto, CA: AAAI.

Granger, R. (2006). Engines of the brain: The computational instruction set of human cognition. *AI Magazine*, 27(2), 15.

Gratch, J., Hartholt, A., Dehghani, M., & Marsella, S. (2013). Virtual humans: A new toolkit for cognitive science research. *Applied Artificial Intelligence, 19*, 215–233.

Gratch, J., Rickel, J., André, E., Cassell, J., Petajan, E., & Badler, N. (2002). Creating interactive virtual humans: Some assembly required. *IEEE Intelligent systems, 17*(4), 54–63.

Hart, D., & Goertzel, B. (2008). Opencog: A software framework for integrative artificial general intelligence. In: *Frontiers in Artificial Intelligence and Applications, Proceeding. 1st AGI Conference* (Vol. 171, pp. 468–472)

Hartholt, A., D. Traum, S. C. Marsella, A. Shapiro, G. Stratou, A. Leuski, L. P. Morency, and J. Gratch. (2013). "All together. now." In *Proceedings of the 13th International Conference on Intelligent Virtual Agents* (pp. 368–381). Berlin, Germany: Springer, August 29–31.

Hilgard, E. R. (1980). The trilogy of mind: Cognition, affection, and conation. *Journal of the History of the Behavioral Sciences, 16*(2), 107–117.

Ho, W. C., & Dautenhahn, K. (2008). Towards a narrative mind: The creation of coherent life stories for believable virtual agents. In H. Prendinger, J. Lester, M. Ishizuka (Eds.), *Intelligent Virtual Agents,* IVA 2008, Lecture Notes in Computer Science (Vol. 5208, pp. 59–72). Berlin, Germany: Springer.

Johnson, T. R. (1997). Control in ACT-R and Soar. Control in Act-R and Soar. In M. Shafto & P. Langley (Eds.), *Proceedings of the Nineteenth Annual Conference of the. Cognitive Science Society* (pp. 343–348). Hillsdale, NJ: Lawrence Erlbaum Associates.

Kiryazov K., & Grinberg M. (2010) Integrating emotions in the TRIPLE ECA model. In A. Esposito, N. Campbell, C. Vogel, A. Hussain, A. Nijholt (Eds.) *Development of Multimodal Interfaces: Active Listening and Synchrony,* Lecture Notes in Computer Science (Vol. 5967, pp. 122–133). Berlin, Germany: Springer.

Klug, M., & Zell, A. (2013). Emotion-based human-robot-interaction. In *Computational Cybernetics (ICCC), 2013 IEEE 9th International Conference on* (pp. 365–368). IEEE.

Kopp, S., Krenn, B., Marsella, S., Marshall, A. N., Pelachaud, C., Pirker, H., Thórisson, K. R., & Vilhjálmsson, H. (2006). Towards a common framework for multimodal generation: The behavior markup language.

In J. Gratch, M. Young, R. Aylett, D. Ballin, P. Olivier (Eds.) *Intelligent Virtual Agents,* IVA 2006, Lecture Notes in Computer Science (Vol. 4133, pp. 205–217). Berlin, Germany: Springer.

Kshirsagar, S., Guye-Vuilleme, A., Kamyab, K., Magnenat-Thalmann, N., Thalmann, D., Mamdani, E. (2002). Avatar markup language. In *Proceedings of 8th Eurographics Workshop on Virtual Environments,* (pp. 169–177). New York: ACM Press.

Laird, J. E. (2012). *The Soar Cognitive Architecture.* Cambridge, MA: MIT press.

Lehman, J. F., Laird, J. E., & Rosenbloom, P. S. (1996). A gentle introduction to Soar, an architecture for human cognition. *Invitation to Cognitive Science, 4*, 212–249.

Lim, M. Y., Dias, J., Aylett, R., & Paiva, A. (2012). Creating adaptive affective autonomous NPCs. *Autonomous Agents and Multi-Agent Systems, 24*(2), 287–311.

Lin, J., Spraragen, M., Blythe, J., & Zyda, M. (2011). EmoCog: Computational integration of emotion and cognitive architecture. In *Proceedings of the Twenty-Fourth.FLAIRS Conference.* Palo Alto, CA: AAAI.

Marinier, R. P., & Laird, J. E. (2007, January). Computational modeling of mood and feeling from emotion. In *Proceedings of the Annual Meeting of the Cognitive Science Society* (Vol. 29, No. 29). Hillsdale, NJ: Lawrence Erlbaum Associates.

Marriott, A. (2001). VHML–Virtual human markup language. In *Talking Head Technology Workshop,* at OzCHI Conference (pp. 252–264). Available online http://ivizlab.sfu.ca/arya/Papers/Others/Representation%20and%20Agent%20Languages/OzCHI-01-VHML.pdf.

Martínez-Miranda, J., Bresó, A., & García-Gómez, J. M. (2012). The construction of a cognitive-emotional module for the help4Mood's virtual agent. *Information and Communication Technologies applied to Mental Health, 34*(3).

Newell, A. (1992). Precis of unified theories of cognition. *Behavioral and Brain Sciences, 15*, 425–492.

Newell, A., & Simon, H. (1956). The logic theory machine–A complex information processing system. *IRE Transactions on Information Theory, 2*(3), 61–79.

Nilsson, N. (1998). *Artificial Intelligence: A New Synthesis.* San Francisco, CA: Morgan Kaufmann.

Ortony, A., Clore, G. L., & Collins, A. (1990). *The Cognitive Structure of Emotions.* Cambridge, MA: Cambridge University Press.

Ritter, S., Anderson, J. R., Koedinger, K. R., & Corbett, A. (2007). Cognitive tutor: applied research in mathematics education. *Psychonomic Bulletin & Review, 14*(2), 249–255.

Rodriguez, L. F., Galvan, F., Ramos, F., Castellanos, E., García, G., & Covarrubias, P. (2010). A cognitive architecture based on neuroscience for the control of virtual 3D human creatures. *International Conference on Brain Informatics* (pp. 328–335).

Russell, J. A., & Mehrabian, A. (1977). Evidence for a three-factor theory of emotions. *Journal of Research in Personality, 11*(3), 273–294.

Scherer, S., Marsella, S., Stratou, G., Xu, Y., Morbini, F., Egan, A., & Morency, L. P. (2012). Perception markup language: Towards a standardized representation of perceived nonverbal behaviors. In Y. Nakano, M. Neff, A. Paiva, M. Walker. (Eds.) *Intelligent Virtual Agents*, LNCS (Vol. 7502, pp. 455–463). Berlin, Germany: Springer.

Slater, S., & Burden, D. (2009). Emotionally responsive robotic avatars as characters in virtual worlds. In *Games and Virtual Worlds for Serious Applications, VS-GAMES'09. Conference in 12–19*. IEEE.

Sloman, A. (2001). Beyond shallow models of emotion. *Cognitive Processing, 2*(1), 177–198.

Sloman, A. (2003). *The Cognition and Affect Project: Architectures, Architecture-Schemas, and the New Science of Mind*. Technical Report School of Computer Science, University of Birmingham, Birmingham, UK. Available online https://pdfs.semanticscholar.org/b376/bcfcd69798a5027e-ae518d001cbaf629deae.pdf.

Smart, P. R., Scutt, T., Sycara, K., & Shadbolt, N. R. (2016). Integrating ACT-R cognitive models with the unity game engine. In J. Turner, M. Nixon, U. Bernardet, & S. DiPaola (Eds.), *Integrating Cognitive Architectures into Virtual Character Design* (pp. 35–64). Hershey, PA: IGI Global.

Smart, P. R., Tang, Y., Stone, P., Sycara, K., Bennati, S., Lebiere, C., Mott, D., Braines, D., & Powell, G. (2014) Socially-distributed cognition and cognitive architectures: Towards an ACT-R-based cognitive social simulation capability. *At Annual Fall Meeting of the International Technology Alliance Annual Fall Meeting of the International Technology Alliance, United Kingdom* (p. 8) September 15, 2014.

Sutskever, I., Brockman, G., Altman, S., & Musk, E. (2016). OpenAI technical goals [Web log post]. Available online https://blog.openai.com/openai-technical-goals/.

Swartout, W. R., Gratch, J., Hill Jr, R. W., Hovy, E., Marsella, S., Rickel, J., & Traum, D. (2006). Toward virtual humans. *AI Magazine, 27*(2), 96.

Tulving, E. (1983). Ecphoric processes in episodic memory. *Philosophical Transactions of the Royal Society of London B, 302*(1110), 361–371.

Ylvisaker, M., Hibbard, M., & Feeney, T. (2006). "What is cognition". Available online LEARNet on The Brain Injury Association of New York State website: http://www.projectlearnet.org/tutorials/cognition.html.

CHAPTER 7

Embodiment

INTRODUCTION

In this text so far, virtual humans have been considered as relatively isolated, self-contained entities. Whilst they may have sensors to detect and actuators or servos to change what is happening beyond their 'selves', little consideration has, so far, been given as to how they relate to the space beyond their avatar, and whether that relationship is actually important. Science fiction is full of disembodied super-intelligences quite content to live on hard drives or distributed through the internet. However, much of the research into intelligence places an emphasis on the concept of embodiment – the importance of a person having a 'physical' sense of self, giving it both a domain to operate within and be influenced by, and a way to 'ground' its experiences. This embodiment is seen by many as a vital part of consciousness. Whilst such embodiment could come in the form of robotics this current work is more interested in the potential role of virtual worlds and social virtual reality in providing an opportunity for embodiment, grounding, and context for virtual humans.

This chapter begins by examining the current and significant research around the concept of embodiment, and some of the challenges to it. The concept and importance of 'grounding' will also be discussed. The possibilities for the use of virtual worlds in providing an environment for embodiment and grounding will then be presented, and its implications for virtual human development evaluated.

FROM SYMBOLIC TO EMBODIED ARTIFICIAL INTELLIGENCE

The early days of research into AI, from the Dartmouth Conference of 1946 onwards, were focused around the concept of symbolic AI – creating computer programs that could 'understand' and manipulate symbols in order to reason – called the cognitivist approach. Anderson (2003) describes cognitivism as 'the hypothesis that the central functions of mind—of thinking—can be accounted for in terms of the manipulation of symbols according to explicit rules'.

The pinnacle of the symbolic approach (later referred to as 'Good Old Fashioned AI' - GOFAI) were systems such as Cyc (Lenat, 1989), which began in 1984 and spent twenty years, and, 'approximately 900 person-years of effort—building a knowledge base that is intended to capture a broad selection of common-sense background knowledge' (Matuszek et al., 2006). By 2006, Matuszek reports that Cyc had 'more than 2.2 million assertions (facts and rules) describing more than 250,000 terms, including nearly 15,000 predicates' (Matuszek et al., 2006) linking, more or less, every concept and thing that one would expect an AI to know. Cyc is still available at http://www.opencyc.org/.

By the 1950s, though, the idea of an AI which had some form of interface to the real world was being studied, and even Alan Turing had considered adding television cameras, microphones and loudspeakers in order to interact with the physical world. Turing also identified that computer development could either focus on relatively abstract programming problems, such as chess, or focus on how a computer could interact with the real world (Brooks, 1991). This divergence between an unembodied and an embodied approach was prevalent through much of the 60s and 70s, and even into the 80s – the embodied studies tending to focus around robotics, and the unembodied approach, at best, examining AIs within very minimal 'block worlds' – just simple 2D and 3D environments made of squares and cubes which bore little or no resemblance to the real world.

EMBODIMENT AND COGNITION

In the late 80s, researchers began to consider again the role of embodiment within AI research. A principle exponent was Rodney Brooks of the Massachusetts Institute of Technology, and his influential 1991 paper, 'Intelligence without Reason', established the parameters for the new 'embodied AI'. Brooks identified that 'most of what people do in their day-to-day lives is not problem-solving or planning, but rather it is a routine activity in a relatively benign, but certainly dynamic, world' (Brooks, 1991). There is no doubt that giving an AI or virtual human the ability to do those common sense, everyday tasks has proven to be more of a challenge than creating a restricted AI that can drive a car, play chess, or mine vast quantities of data. Brooks asked whether there can actually be such a thing as a disembodied mind, given how much of 'what is human about us is very directly related to our physical experiences' (*Brooks*, 1991).

A major concern that this new generation of AI researchers had was that much of the then current AI research took place in the very simplistic block worlds mentioned earlier – worlds which were uniquely created for the AI, and which lacked any sort of richness, or real similarity to the real world. Summed up in his phrase '[t]he world is its own best model', Brooks (1991, 2) pointed out that:

> In order to really test ideas of intelligence, it is important to build complete agents which operate in dynamic environments using real sensors. Internal world models which are complete representations of the external environment, besides being impossible to obtain, are not at all necessary for agents to act in a competent manner.

Brooks identified four key aspects of this new approach to AI.

- **Intelligence** – the intelligence comes not just from computation but also from the coupling of the robot with the world, i.e., its behaviours in the world.
- **Emergence** – the intelligence also emerges from the robot's interactions with the world, and between its components, with no one part of the system necessarily being the cause.
- **Situatedness** – the robots are situated in the real world, not dealing with abstract descriptions.
- **Embodiment** – the robots have bodies and experience the world directly, their actions are part of a dynamic with the world and have immediate feedback on their own sensations.

Each of these will now be considered in turn.

Intelligence and Emergence

For many people, it is the interaction between the world and the individual (whether human, animal, robot or virtual human) which gives meaning to intelligence; without such interaction, 'intelligence' is almost meaningless.

Where previously cognitivist researchers had talked about perception, planning, modelling, and learning sub-systems which controlled the behaviour of the AI, this new generation saw the opposite causality; it was these activities that emerged from the behaviours.

Situatedness

Situatedness means that the individual (or AI or virtual human) is operating within a 'real' world. They are not operating in a totally artificial environment created solely for the purposes of developing an AI. To create a true AI, it needs to inhabit a real world with all the chaos, uncertainty, unpredictability and scope of action that that brings. In addition, the AI should have no privileged access to information about the world, and, ideally, it should gain such information in the same way as the human inhabitants, otherwise, the boundary between the AI and the world becomes blurred.

Dreyfus argues that 'the meaningful objects ... among which we live are not a model of the world stored in our mind or brain; they are the world itself' (Dreyfus, 2007), echoing Brooks' statement about the world as its own best model made earlier. So, by placing the AI in a real environment, not only is the need to create an artificial one to fully test the AI avoided, but the AI can, like us, use the world (or potentially a virtual world designed for humans, not specifically for AIs) as its test-bed from which to learn.

Embodiment

If situatedness is about operating in a real world, then embodiment is about having a body in that world. A virtual human represented solely by a 2D head-and-shoulders avatar is unlikely to be able to develop the same level of humanness, intelligence, or even self-awareness as one which has a full body presence – be that robotic or virtual.

Zeimke (2001) identified five different types of embodiment, adding a sixth in 2003.

- **Structural coupling** between agent and environment, i.e., the agent can influence the environment, and the environment can influence the agent.
- **Historical embodiment**, i.e., 'the embodiment is, in fact, a result or reflection of a history of agent-environment' – which could even apply to how language reflects the embodied existence.
- **Physical embodiment**, the agent needs to have a physical instantiation.
- **Organismoid embodiment**, i.e., organism-like bodies which at least to some degree have the same or similar form and sensorimotor capacities as living bodies.
- **Organismic embodiment** of autonomous and autopoietic (i.e., self-reproducing and maintaining) living systems.
- **Social embodiment**, the extent to which states of the body, such as postures, arm movements and facial expressions, arise during social interaction and play central roles in social information processing.

It should be noted that whereas the first five represent an increasingly high set of hurdles for a system to jump over in order to be deemed embodied, the sixth represents a different take on the notion of embodiment.

Lakoff and Johnson (1999) identify the following similar elements to embodiment, but with less of an emphasis on the organismic aspects.

- Physiology (biological design affecting how things sensed/perceived).
- Evolutionary history (e.g., how metaphors are informed by embodiment).
- Practical activity and learning by trial and error.
- Socio-cultural situatedness.

GROUNDING

The 'symbol grounding problem' is a significant issue in cognitive science. At its heart, it is about how the 'symbols' that a human or virtual human might manipulate (e.g., the words 'table' or 'zebra') can be given real meaning rather than just being defined by other symbols. As Harnad puts it '[h]ow is symbol meaning to be grounded in something other than just more meaningless symbols' (Harnad, 1990).

The ways that embodiment provides a solution to the grounding problem is one that attracted the attention of AI researchers. As Brooks (1991, 15) says:

> There are two reasons that embodiment of intelligent systems is critical. First, only an embodied intelligent agent is fully validated as one that can deal with the real world. Second, only through a physical grounding can any internal symbolic or other system and a place to bottom out, and give 'meaning' to the processing going on within the system.

Whilst the first appears to refer back to situatedness, it is this notion of 'grounding' that has usually been seen as the central thread of embodiment. Anderson (2003, 92) offers:

> In my view, it is the centrality of the physical grounding project that differentiates research in embodied cognition from research in situated cognition, although it is obvious that these two research programs are complementary and closely related.

As an example of what is meant by 'grounding', consider a chair. A person who has never seen a chair before could be given a vast amount of data about what a chair is, its purpose, how it's made, the different styles of chairs, images and video of chairs in use, equations for the mechanics of how the chair supports a person's weight and how it relieves the weight on their legs. But until such time as they actually sit in that chair and *feel* the weight lift from their legs, perhaps sink into the deep upholstery of the chair, and all this after a long walk or a hard day at the office, they don't really *know* what a chair is. They can then extend that concept of the chair to something physically un-chair like, but which provides the same sense of relaxation when used as a chair – such as a tree-stump (Anderson, 2003). This seems to be a very similar issue to that of 'qualia', the ineffable subjective properties of experiences, as described in Jackson's (1986) thought-experiment of Mary, a colour scientist who has lived her whole life in a black and white world, and the way that her understanding of colour changes the moment she first sees a red tomato. By giving a virtual human embodiment, it can ground all those symbols that it has been taught, it can know what a chair is for, what else can pass for a chair, the ways in which a chair could be abused (for example, as a step to reach something), and, ultimately, what it feels like to use one.

An embodied AI, which is not situated, is realizable – in fact, it is an extension of exactly the sort of situation that Brooks was complaining against – an AI embodied in a robot body but then placed with a completely artificial environment designed only to develop and test the capabilities of the AI. Conversely, a situated AI, which is not embodied, is also realisable – consider something like an advanced version of Siri which has to deal with the visual and audio clutter of the real world – but which is not really embodied. The interesting developments come when the AI or virtual human is both situated and embodied, and able to explore and interact with its (our) world on its own terms. As Haugeland (1998) says, 'cognition is embedded and embodied'.

ENACTIVE AI

The works of Heidegger (1958) on existentialism, and of Merleau-Ponty (1962) on the phenomenology of the body have influenced a further refinement to the embodied AI model called Enactive AI. The most influential work in this area is *The Embodied Mind* by Varela, Thompson and Rosch (Varela et al., 1991). The book discusses embodiment in a double sense, that of the self as both a physical structure and as a 'lived, experiential' structure, and presents an 'enactive' account of perception. Enactive AI is defined by two elements. First, that 'perception consists in perpetually guided action' – that is, the self 'both initiates and is shaped by the environment'. Second, that 'cognitive structures emerge from the recurrent sensorimotor patterns that enable action to be perceptually guided'. This element is informed by the work of Piaget (1954) and focuses on the way that the self builds (and shares) the categories, relationships and models (the structures) it needs to understand the world through its exploration and direct experience and manipulation of the world.

Varela et al. (1991, 205) summarise enactive cognition as providing:

> A view of cognitive capacities as inextricably linked to histories that are lived, much like paths that exist only as they are laid down in walking. Consequently, cognition is no longer seen as problem-solving on the basis of representations; instead cognition in its

> most encompassing sense consists in the enactment or bringing forth of a world by a viable history of structural coupling.

Froese and Ziemke (2009) describe Enactive AI as complementing and refining embodied AI, balancing the behaviour focus of traditional embodiment with a new focus on the self-identity of the entity which creates that behaviour, and that meaning is an aspect of the relational domain established between environment and the agent.

Enactive AI also has an additional focus on emergence. Froese and Ziemke (2009, 485) argue:

> Whereas embodied AI is faced with the challenge of designing an agent such that, when it is coupled with its environment, it gives rise to the desired behavior, here the target behavioral phenomenon is one step further removed. We need to start thinking about how to design an environment such that it gives rise to an agent which, when it is coupled with its environment, gives rise to the desired kind of behavior.

This is the so-called hard problem of Enactive AI. Embodied AI is being by being, whereas Enactive AI is being by doing. Whilst some of the emergent and autopoietic elements of Enactive AI may be hard to implement, the focus on 'being by doing' gives some useful direction beyond just having a virtual human embodied within the world, and highlights how the environment could, and should, shape the virtual human.

CHALLENGES TO EMBODIED AI

Whilst Embodied AI has undoubtedly been a driving force in AI development over the past few decades, it is not without its detractors. Just as the Embodied AI fraternity referred to the Symbolic AI tradition as 'Good Old Fashioned AI' – GOFAI, so the challengers of embodiment referred to it as 'Nouvelle AI' (Sloman, 2009), one assumes in the sceptical manner of *nouvelle cuisine* or *nouvelle vague* cinema. Sloman's challenge is that much of human existence and thought is *not* concerned with the physical world, the world of embodiment and situatedness. 'The notion that intelligence involves constant interaction with the environment ignores most of human nature' (Sloman, 2009, 253). Examples include much pure mathematics, philosophy, music and abstract art. For Sloman, it is a combination of symbolic and embodied approaches that are perhaps the best way forward.

An alternative challenge comes from Dreyfus (2007) in his discussion on Heideggerian AI. He calls for a focus on the end result – the active, embodied being, 'being-in-the-world' in Heideggerian thought, rather than in the constituent, often artificial, techniques used to achieve it. Dreyfus identifies Freeman's Merleau-Pontian neurodynamics (Freeman, 1991) as a possible solution. In this, an animal's perceptual system is primed by past experience and arousal to seek and be rewarded by relevant experiences. Dreyfus (2007, 1153) suggests:

> Thus, according to Freeman's model, when hungry, frightened, etc., the rabbit sniffs around seeking food, runs toward a hiding place, or does whatever else prior experience has taught it is successful. The weights on the animal's neural connections are then changed on the basis of the quality of its resulting experience. That is, they are changed in a way that reflects the extent to which the result satisfied the animal's current need.

This certainly reflects the neural-net/deep-learning type approaches that many are using in developing AI.

Dreyfus (2007, 1154) goes on to highlight that there is 'no fixed and independent intentional structure in the brain – not even a latent one', and that the brain state 'only comes into existence and only is maintained as long as, and in so far as, it is dynamically coupled with the significant situation in the world that selected it and does not exist apart from it' (Dreyfus 2007, 1154). Perhaps, this begins to identify the difference between a self-aware, sentient entity, and an entity which is not a self-aware entity. In the rabbit's case, its lack of self-awareness or inner narrative means that its brain states only come into existence as needed by its interactions with the world, whereas in a self-aware, sentient, entity the inner narrative requires an on-going series of brain-states.

VIRTUAL HUMANS AND VIRTUAL WORLDS

Most of the discussion about situatedness and embodiment has been in the context of AI and physical robots. As discussed above, one of the drivers for an embodied and situated approach to AI was to move away from artificial, virtual, block worlds, to the real, everyday, physical environment so that the AI could not only ground itself and its concepts but could also learn to deal with the clutter and chaos of the real world.

However, Social Virtual Worlds, such as Second Life (www.secondlife.com), offer the opportunity to explore how the ideas of embodiment and situatedness could apply to virtual humans in virtual worlds. These 3D digital multi-user virtual environments (MUVEs) are used by humans (through their avatars) for a whole variety of business, recreational and academic purposes – not just for playing a single game, and are, as almost any resident would attest, pretty chaotic.

The main challenge that could be made to virtual humans, in the form of an avatar in a virtual world, being 'embodied' is that, unlike an AI driving a physical robot, they have no physical bodies – they literally have no physical embodiment. However, whilst a robot may have a physical body, it may have no more sense of embodiment than an avatar. The robot 'brain' has a set of connections out to its limbs. These pass sensory information (touch, resistance, strain, level of effort, etc.) up from the limbs to the brain and take commands to move from the brain to the limbs. The brain is just dealing with a set of sensory inputs and a repertoire of commands. It doesn't have any 'special' sense of the limb – it's just an actuator. If the robot needs to feel 'tiredness', then it needs to be expressed in terms of how much work it has done, how much strain it is feeling from the actions, how fast it is using energy and what energy reserves it has. It won't feel tiredness in a physiological way – but these inputs could be expressed by the robot in more human terms such as tiredness or pain. It may have the ring of Marvin saying that he's got a 'terrible pain in all the diodes down my left side' – but it probably is little more than that.

For a virtual human controlling a virtual avatar, there may not be the same range of sensations or repertoire of actions as a human has, but it has, fundamentally, the same set of inputs and outputs. It would certainly be possible to take the software 'skeleton' that controls the animation of the avatar and use that to provide the virtual human with more of the proprioceptive/kinesthetic sense of where its limbs are that a real human has, just in the same way as a robot would also need to have that proprioceptive sense recreated. The same parallels exist when considering how a robot and an avatar stand in terms of the grounding of symbols. A robot may see a chair and, knowing that it will reduce its energy needs, it may choose to sit in the chair. Whether it then develops a sense of satisfaction from relaxing in a nice comfy chair is moot. But the same goes for an avatar. Now, in real terms, it has no lesser energy requirements when sitting than when standing, but there is no reason why it should not be given an energy budget, which it needs to build up through some activities (plugging in, eating), and which it continually expends in other activities. The relief provided by the chair will then be no less real than for the robot – they are both managing an energy budget.

It should be noted that, here, the discussion is very much about creating a virtual human and introducing similar limitations and drives in the virtual world for the virtual human that physical human faces in the physical world. If the concern is to develop a more idealized AI, then perhaps such limitations can be removed – but one of the key tenets of embodied AI is that the AI needs those limitations in order to make sense of the world *we* live in.

A counter challenge as to whether an avatar is embodied is to consider avatars being driven by physical humans, rather than virtual humans. Whilst the physical human is likely to remain aware of their physical surroundings (although several studies, e.g., Agarwal and Karahanna (2000), show that when computer users – particular gamers in 3D worlds – are in the 'flow', then they have a minimal sense of the temporal or physical environment), it is the digital environments, and their avatar's place and behaviour within it, which will be giving them their sense of embodiment. Yes, they bring the grounding of symbols across to the virtual world (and, in fact, how well those transfers work is probably a good measure of the realism of the virtual world), but they very much feel situated and embodied in the virtual world, as could a virtual human.

Assessing Virtual Worlds against Brooks' Five Principles

Another way of considering the issue of embodiment and situatedness in virtual humans embedded within virtual worlds is to try and apply some of the definitions given above to the virtual realm. As discussed, Brooks (1991) described five principles for his 'new' work on AI. Consider each of these when applied to a virtual realm:

- *'The goal is to study complete integrated intelligent autonomous agents'.*

  A virtual human, represented by an avatar in a virtual world, is also a complete integrated intelligent autonomous agent.

- *'The agents should be embodied as mobile robots, situated in unmodified worlds found around our laboratory. This confronts the embodiment issue'.*

  This statement is equally valid if an avatar is substituted for robot. A social virtual world, such as Second Life, designed quite explicitly for humans not bots (in fact, bots were explicitly banned during the early history of Second Life), is exactly an unmodified world far larger than any laboratory – and indeed with the vast majority of it outside of the researcher's control.

- *'The environments chosen are for convenience, although we strongly resist the temptation to change the environments in any way for the robots'.*

  Again, this statement stands since the researcher can choose spaces within the Social Virtual World which are the most convenient, but need not, indeed, should not (and usually cannot), change them in order to make them more virtual human-friendly.

- *'The robots should operate equally well when visitors, or cleaners, walk through their workspace, when furniture is rearranged when lighting or other environmental conditions change, and when their sensors and actuators drift in calibration. This confronts the situatedness issue'.*

  Taken in respect of avatars not robots, again, this statement stands as it is. In a virtual world such as Second Life, other peoples' avatars often wander and fly through your own land, and the virtual human needs to cope with that. Indeed, it can be useful to make the location of the bot attractive to human users in order to increase human football to continually engage and test the bot. Given the right permissions, those visitors can also make an object appear (called 'rezzing') and move objects. Whilst lighting *per se* is a data issue to the avatar rather than a visual one, there are certainly other environmental conditions that can change which the avatar may have to cope with, such as whole areas suddenly going out of bounds, or the whole world rebooting!

- *'The robots should operate on timescales commensurate with the timescales used by humans. This too confronts the situatedness issue'.*

  Since the whole of the virtual world operates at a human timescale, then, by default, the virtual human needs to operate on that, too. In fact, there are occasions when the virtual human is deliberately slowed in order to seem more human (e.g., reducing typing speeds).

So, apart from changing robot to avatar, every one of Brooks' guiding principles can apply to research with virtual humans as avatars in social virtual worlds.

Assessing Virtual Worlds against Other Embodied Models

Considering some of the other definitions given above, Dreyfus's (2007) description of Merleau-Ponty, Gibson, and Freeman taking 'as basic that the brain is embodied in an animal moving in the environment to satisfy its needs' can also be directly applied to a virtual human operating as an avatar in a virtual world when that virtual human has some established set of motivational needs.

Freeman's model based on Merleau-Ponty's neurodynamics model (1991) could also be directly applied to the virtual human (or virtual rabbit). Social Virtual Worlds can provide a rich ecosystem to operate in, particularly if they also contain elements of artificial life (for example, the Svarga region reported in Calongne and Hiles, 2007), and so an (animal or human) avatar's perceptual system being 'primed by past experience and arousal to seek and be rewarded by relevant experiences' is perfectly implementable. In fact, as previously discussed, development can focus on the implementation of the software to create this, rather than on the physically complex mechatronics of creating the robot and its associated interactions with the physical environment.

Finally, the enactive model needs to be considered. This presents probably the greatest challenge to the virtual human/avatar model, particularly in terms of the autopoietic element which is present in its most developed versions. However virtual worlds themselves may be just what the enactive model needs. As the Enactivists say, the aim should be to focus on creating an environment which gives rise to an agent with the desired behaviour, not starting by creating the agent. There are no reasons why social virtual worlds should not be such environments – in fact, they are probably more suitable than the physical world which has very little scope for redesign. The AI that could emerge would probably not be a virtual human, but it would be of considerable interest all the same!

CONCLUSION

This chapter has examined how AI research has moved from purely symbolic approaches to cognition and development to approaches driven by the ideas of embodiment and enactivism. By embodying the virtual human, it gains the ability to ground the symbols it is manipulating, and its intelligence and cognitive capability are expressed and shaped through its ability to interact with, explore and shape the virtual world it operates in. Whilst early AI researchers were dismissive of digital 'block-worlds' and saw physical robotics as the way to embody AIs in a rich, chaotic and uncontrolled environment, this chapter has shown how virtual worlds, particularly social virtual worlds, offer, perhaps, the ideal environment in which to embody and develop virtual humans – offering all the richness of a human environment but without all the complexities of building and operating physical robots.

Having examined the various procedural, architectural and conceptual elements that can go into creating a virtual human, the next chapter will look at how they can all be brought together to create a virtual human, the roles that such a virtual human might undertake and whether these elements alone are actually sufficient to create a virtual human.

REFERENCES

Agarwal, R., & Karahanna, E. (2000). Time flies when you're having fun: Cognitive absorption and beliefs about information technology usage. *MIS Quarterly, 24*(4), 665–694.

Anderson, M. L. (2003). Embodied cognition: A field guide. *Artificial intelligence, 149*(1), 91–130.

Brooks, R. A. (1991). Intelligence without reason. In: R. Chrisley & S. Begeer (Eds) *Artificial intelligence: Critical Concepts, 3*, 107–163. Cambridge, MA: MIT Press.

Calongne, C., & Hiles, J. (2007). Blended realities: A virtual tour of education in second life. In *TCC Worldwide Online Conference* (pp. 70–90). TCC Hawaii.

Dreyfus, H. L. (2007). Why Heideggerian AI failed and how fixing it would require making it more Heideggerian, *Artificial Intelligence, 171*(18), 1137–1160.

Freeman, W. J. (1991). The physiology of perception. *Scientific American, 264*(2), 78–87.

Froese T., & Ziemke T. (2009). Enactive artificial intelligence: Investigating the systemic organization of life and mind. *Artificial Intelligence 173*(3–4), 466–500.

Harnad, S. (1990). The symbol grounding problem. *Physica D: Nonlinear Phenomena, 42*(1–3), 335–346.

Haugeland, J. (1998). 'Mind embodied and embedded,' In J. Haugeland (Ed). *Having Thought: Essays in the Metaphysics of Mind* (207–237). Cambridge, MA: Harvard University Press.

Heidegger, M. (1958). *The Question of Being.* Lanham, MD: Rowman & Littlefield.

Jackson, F. (1986). What Mary didn't know. *The Journal of Philosophy, 83*(5), 291–295.

Lakoff, G., & Johnson, M. (1999). *Philosophy in the Flesh* (Vol. 4). New York: Basic Books.

Lenat, D., & Guha, R. V. (1989). *Building Large Knowledge-Based Systems: Representation and Inference in the Cyc project*. Reading, MA: Addison-Wesley Publishing Company.

Matuszek, C., Cabral, J., Witbrock, M. J., & DeOliveira, J. (2006). An introduction to the syntax and content of cyc. In *AAAI Spring Symposium: Formalizing and Compiling Background Knowledge and Its Applications to Knowledge Representation and Question Answering* (pp. 44–49). Palo Alto, CA: AAAI.

Merleau-Ponty, M. (1962). *Phenomenology of Perception*. London, UK: Routledge.

Piaget, J. (1954). *The Construction of Reality in the Child*. London, UK: Routledge.

Sloman, A. (2009). 'Some requirements for human-like robots: Why the recent over-emphasis on embodiment has held up progress' In B. Sendhoff et al. (Eds.), *Creating Brain-Like Intelligence,* LNAI (Vol. 5436, 248–277). Berlin, Germany: Springer-Verlag.

Varela, F. J., Thompson, E., & Rosch, E. (1991). *The Embodied Mind: Cognitive Science and Human Experience*. Cambridge, MA: MIT Press.

Zeimke, T. (2001). Are robots embodied? In *Proceedings of the First Intl. Workshop on Epigenetic Robotic* (Vol. 85). Lund, Sweden: Lund University Cognitive Studies.

CHAPTER 8

Assembling and Assemblages

In Chapters 3 through 7, this book has examined the different technologies and approaches that can contribute in making the avatar body, mind, and senses of a virtual human, and possible architectures to integrate those, and how the resultant virtual human can be grounded through embodiment within a virtual world. But how can these come together to create a virtual human, and what roles might virtual humans fulfil? Even when presented with a computer system masquerading as a human, there is still a sense that assembling a believable virtual human may require more than just the individual elements presented, and that the virtual human needs, in some ways, to be more than the sum of its parts.

This chapter will first examine a possible sequence of steps to build a virtual human. An example of a virtual human under development will then be considered before describing some typical roles that a virtual human might fulfil at increasing levels of sophistication. Finally, the concept of a virtual human will be considered through a number of different lenses.

BUILDING A VIRTUAL HUMAN

There are many and varied approaches to building a virtual human. The early architectures described in Chapter 6 started with the building of a problem-solving system and then added to that the elements (such as natural language and emotions) that were needed to create a more rounded and capable virtual human. Another approach (which dates back to at least Turing) is to try and mirror human development, creating a virtual baby or virtual toddler and then having that learn in the step-by-step way that human babies and toddlers do. However, this linear step-by-step approach is under debate (MacBlain, 2018). In contrast, many more recent systems have taken the natural language elements as a base (as described in Chapter 5), and then added the decision-making and other functionality required. Much current attention is based around machine learning approaches, but these require a decent training set of data and examples with which to work, and whilst it may be a useful technique for specific areas of functionality (such as speech and pattern recognition), it may be less well suited to the management or control of the overall system.

The approach described below, which is that used by the authors, largely follows the chatbot-centric approach.

The Natural Language Core

Given the chosen approach, the heart of the example virtual human will be some form of natural language system. This could be based on the intention approach used by systems such as Alexa and Siri, but a more open conversational approach is required to give a real sense of humanness – so, some form of chatbot engine, or, possibly, a machine-learning approach. Even in choosing a chatbot engine, there is a need to move away from the question and answer model that is typical in chatbots but atypical in human conversation where the turn taking is far more asymmetrical. A useful feature of the chatbot approach is that the same mechanisms that take language input and then decide on the appropriate language output can also be used to take environmental sensory input (for instance, object approaching) and decide on the appropriate corporeal response (for example, raise eyebrows, shut, run away).

Memory and Knowledge

Whilst there is often a temptation to embed the knowledge of a chatbot within the chatbot scripts and language-based data structures, this is best avoided – as it makes it almost impossible for the chatbot to learn. Instead, database systems can be used to capture the semantic and procedural knowledge that the chatbot needs – and, as described in Chapter 4, a semantic triples approach can provide a very flexible and human-like knowledge store. As the chatbot, or virtual human, begins to interact with other people, then new memories and knowledge will be generated. In a commercial environment, this may be about user needs and preferences and be stored in a conventional commercial database. In a social virtual world environment, this may be more human-like episodic memories of people and encounters.

As memory builds up, there may be a need to manage it – to remove memories that are no longer useful, or keep important memories close at hand. In the commercial sphere, this may mean backing-up and archiving memory. In a more human-like system, it may mean giving the virtual human some sleep time to cull memories that are too similar to the pattern of normal life and just keeping those with a high emotional valence. The churn of this processing could even be experienced as dreaming.

Problem-Solving and Decision-Making

In commercial virtual humans, there will be a lot of 'decision-making' required (what service does this user need, what information do I need from them, which is the best solution, what is the likelihood of upselling?), but this is likely to be highly parameterised. It may be that the virtual human only acts as the front end to a larger commercial Customer Relationship Management (CRM) or Enterprise Resource Planning (ERP) system, collecting the required data and having the backend system advise on next steps, as required. A more social or autonomous virtual human will need to be able to deal with a wider variety of situations, and so will need a more general-purpose problem-solving approach (such as the SOAR and ACT-R architectures described in Chapter 6). Both will require some ability to 'learn' what the best approaches are to given problems, with, again, a wider scope being required by the social/autonomous AI.

Representation

So far, the only representation that this prototype virtual human needs to have is a text-chat interface – and, indeed, for many commercial, and some social virtual humans, this is all that is used. But to better present as human, it is relatively trivial to add an animated head-and-shoulders or even full-body avatar to the virtual human. Chapter 3 described the levels of detail that can now be obtained when creating a digital avatar – if such detail is needed for the role – and having an avatar that looks no different from a physical human (and most certainly over a video-conferencing link) is not at all far off. Adding text-to-speech to give it a voice is also relatively trivial. Speech recognition a bit harder, but as Google has shown with Google

Duplex (Leviathan, 2018), such voices and supporting natural language capability are already good enough to be mistaken for humans in some situations (such as booking a restaurant table in the Google case).

So far, the prototype virtual human being described is probably very 'robotic', it is more virtual humanoid than virtual human. It can look human, have a conversation, remember things and make decisions and solve problems, but it is lacking several features that many people will see as important for humanness.

Emotions and Motivation

Adding an emotional architecture to the virtual human can create far more believable behaviours, and in a very adaptable way. The virtual human will start to appraise what is happening, and decide on the appropriate response and then show them – possibly with a 'low-road' instinctive reaction and a more measured 'high-road' response. When encountering a new situation or stimuli, it will assesses what the effect is on itself, and then assign the appropriate emotional response (for example, happiness, fear, etc.) if encountered again.

To give more meaning to the emotional responses they need to be linked to a motivational model. The virtual human needs a reason to 'be', and a reason to undertake almost everything that it does, and a reason that goes beyond being explicitly being told (or programmed to do it). A simple model, such as Maslow's Triangle discussed in Chapter 4, can offer the prototype virtual human with a set of multi-level (and often competing) goals, drives and motivations, which can result in complex and seemingly unpredictable and emergent behaviour.

Embodiment

Whilst the prototype virtual human is confined to a 1D text-based or 2D avatar-in-a-window existence, it is unlikely to develop a rich and complex life or personality. As Chapter 7 described, embodying the prototype virtual human as a 3D avatar within a public 3D social virtual world is readily achievable. This not only lets the virtual human ground its semantic knowledge and experiences, but also allows the virtual human to engage with physical humans (and other virtual humans) on a potentially level playing field. The only immediate significant technical challenge is the navigation within, and interaction with, the environment, but as will be described below, these are readily achievable. As a 3D avatar, there is also more scope for creating a human-like avatar, as described in Chapter 3, and more need to provide gesture control and a proprioceptive sense. A physiological model and interoceptive sense add further to the realism – both for the virtual human and the humans it interacts with.

Meta-Management and Personality

In elements already described, from natural language to emotion and gesture, the 'personality' of the prototype virtual human will be present. The real trick is to move away from one which is programmed through a set of parameters to one that evolves as the virtual human has experiences. If the virtual human is being asked to do tasks which don't satisfy any of its needs or drives (i.e., boring tasks), then it will feed sad, which will lead to a depressed mood, which, if continued for a long time, may begin to shift its personality to a more depressed or resigned one – or potentially an angry one. This autonomy of personality (building on an autonomy of motivation) immediately begins to create a virtual human with a degree of independence, but which could make it unattractive for many commercial applications – who needs a bored, passive aggressive virtual customer service agent when human ones are all too readily available!

Another area of long-term development is in the ability of the virtual human to learn about how it itself works, what it likes and doesn't like, what approaches and actions work and what don't, and begin to shape itself, its activities and, maybe, its world accordingly. Here, the virtual human may begin to show a more truly human form of learning, not just learning to solve problems to adapt to small-scale tactical situations but beginning to adopt a life-long approach to learning and adaption.

Its Own Life

As suggested in Chapter 5, the point at which people may accept a virtual human as fully warranting the term human is when it can argue its own case and show an interior life and a continuous internal dialogue and narrative self. Whilst this may seem a long way off, the basic ingredients of such capability are a natural growth of what has already been described, a natural language capability (including conversations with itself, not just with others), a growing collection of episodic memories, which are semantically linked to people and places and, more importantly, to emotions, and a set of long-term but evolving motivations and desires that drive the virtual human forward. There is no suggestion that this prototype virtual human has now achieved any form of consciousness or sentience, and it may well still be far short of the sort of Artificial General Intelligence

(AGI) described in Chapter 13, but it is a long way from a Siri, Alexa or other current virtual agent. And the technology to achieve it, at least to a first level of sophistication, is available today.

HALO – A CASE STUDY

In order to understand the type of process described above, and, in particular, the ideas of embodiment of a virtual human with a virtual world, one of the authors has been working on a project to achieve a chatbot-based virtual human embodied within a social virtual world.

Halo was initially a web-based chatbot developed using a Sitepal animated 2D head-and-shoulders avatar and a modified AIML chatbot engine (see Figure 8.1a and as described in Chapters 3 and 5, respectively) and has been a test-bed at Daden Limited for chatbot and virtual human development for over 10 years.

In 2006, she made her first appearance within the Second Life virtual world. Initially, this was as a static head-and-shoulders avatar on a 'fake' computer screen, but by 2008, she had an avatar body, which looked no different to any other (human-controlled) avatar in Second Life (see Figure 8.1b). The avatar was controlled, at a low level, by the LibOpenMetaverse (LibOMV – https://github.com/openmetaversefoundation/libopenmetaverse) application (essentially a reverse-engineered Second Life client for computer rather than human control), with bespoke C# code providing some of the higher functions and the link to the same AIML derived chatbot engine. This system was described in Slater and Burden (2009) and formed the system that was a finalist in the British Computer Society's (BCS) Specialist Group on Artificial Intelligence's Machine Intelligence competition.

(a) (b)

FIGURE 8.1 Representations of Halo. (a) Halo as a 2D avatar on the web. (b) Halo as a 3D avatar in Second Life.

Within Second Life, Halo is an avatar which looks and moves no differently from a human-controlled avatar and can change clothes, and even appearance, at will – but in the same way as human avatars can. The controlling software (Halo's brain or mind) is able to trigger gestures and facial animations (for example, smiling), again, in the same way as for a human-controlled avatar. Halo has a very simple physiological model based on Second Life's damage system, so that she knows if she's been hurt. Her awareness of her environment is provided not by sight but a reading of the computer data which is being provided to the client software to render the scene visually for a human user (as in *The Matrix*). Halo can communicate with other avatars (both humans and other bots) in natural language through text chat – the traditional communication mode for Second Life users (many still rejecting voice) and can carry out speech conversations when accessed through a 2D web interface. She also has some spatial awareness of the people around her, so she can isolate chat directed at her rather than other people.

Halo's brain incorporates an implementation of the Emotion-AI (E-AI) emotional architecture (Slater and Burden, 2009), which drives her attention, appraisal reaction and emotion systems, and has a mood engine similar in behaviour to the PAD-Space system described in Chapter 6. There is also a motivation engine based on Maslow's hierarchy of needs and similar to the FAtiMA/PSI model also described in Chapter 6. Halo's goal and planning system for immediate and daily tasks and activities is currently less well developed but is slowly evolving. Her memory model is split between a working memory and a long-term memory containing semantic, procedural and episodic components (as in SOAR and OpenCog) and implemented using semantic triples, and with the ability to forget things over time (as in EmoCOG). Halo can learn new information explicitly by questioning, and implicitly by experience (for instance, learning what hurts her and what rewards her). The architecture makes use of the Avatar Sensing Markup Language (ASML) and Avatar Action Markup Language (AAML) noted in Chapter 6 to enable her 'brain' and 'avatar' to be clearly separated and enabling her to be embodied as an avatar in other virtual worlds, other computer systems and, potentially, as a physical robot.

Halo does not have any sense of hearing (for noise or speech) – but this may be rectified given the increased use of voice communications in-world and the increasing ability of speech recognition systems – or any sense of smell, taste or touch (beyond collisions), which are not available to human users within Second Life either. What she also currently lacks, but the team is working on, include a more detailed physiological model, a more sophisticated goal, planning and problem-solving system, possibly based on the SOAR or BDI models, and a meta-management and self-monitoring process. In the longer term, it is planned to give Halo the ability to 'dream' as the daily memory management routines are run during 'sleep' – perhaps with the dreams themselves realized as environments with Second Life or another virtual world, and, ultimately, to enable the generation of an internal narrative. A video of the capabilities demonstrated at the BCS competition is at www.virtualhumans.ai.

There are anecdotal cases of Halo interacting with human-controlled avatars for several minutes before they realize that they are talking to a computer, and as discussed in Chapter 5 an avatar based on Halo's architecture successfully passed a modified covert Turing Test in Second Life. Halo represents an on-going experiment to explore the idea of an embodied virtual human, operating within a social virtual world, which offers a wide variety of opportunities for interaction and learning in situations beyond the developers' and researchers' control. Her development is still, after almost 10 years, at a very early stage, and currently her behaviour is certainly more 'programmed' than would be ideal, but she is already good enough to fool some people, and, in time, may not only fool many more but will take on an existence within Second Life or some other social virtual world, of more or less her own making.

THE ROLES OF VIRTUAL HUMANS

This section will consider how a virtual human might be manifest in a number of different roles, showing a steady increase in expected sophistication. The examples chosen are:

- Intelligent Speaker
- Customer Service Chatbot
- Virtual Assistant
- Virtual Tutor
- Virtual Non-Player Character (NPC)
- Virtual Life Coach
- Virtual Expert
- Virtual Persona

- Virtual Person
- Digital Immortal

Each of these is considered in more detail below.

Intelligent Speaker

Intelligent speaker apps include systems such as Alexa and Google Home. The 'virtual human' element is limited to the natural language capability. There is no 'body' – other than that of the speaker casing, and no emotional or motivational elements. Such reasoning and memory capabilities, as there are, purely support the ability to service the user, such as maintaining preferences and past behaviour histories in order to improve future service. Figure 8.2 shows the Intelligent Speaker in terms of the virtual human traits profile presented in Chapter 1.

Customer Service Chatbot

Customer service chatbots are, along with phone-based personal assistants and smart speakers, one of the most commonly encountered virtual humanoids. Whilst the majority are text-chat driven with no associated imagery, many have static images to represent the brand persona, and some use head-and-shoulders full body avatars with lip-sync'd speech-to-text. The greater reliability of text-chat as against speech means that more complex and extended conversations can be maintained – but also

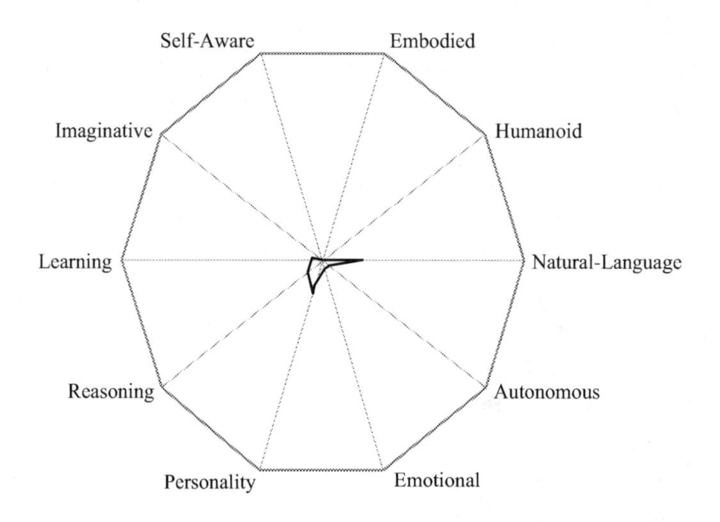

FIGURE 8.2 Intelligent Speaker profile.

that the bot can make use of multiple-choice panels to better manage the interaction, but, perhaps, destroy the illusion of a conversation. Such bots can also be used to front online chat pop-up windows on websites, possibly masquerading as a human and then handing the call off to a physical human if things get too complex. It is possible that a bit more personality may be shown – and enabled through the longer conversational style – but otherwise the customer service chatbot is usually very much in the same virtual humanoid category as smart speakers and most personal assistants.

Personal Virtual Assistant

Personal Virtual Assistants, such as Siri and Cortana, begin to blend some of the capabilities and characteristics of the smart speaker and customer service chatbot, although the focus on voice conversation tends to keep conversations shorter and more task orientated. Like smart speakers, the natural language interface is typically intention-based, with the assistant identifying the user intent (for example, to book a train ticket) and then filling the slots of information required to achieve that goal. Manufacturers' aspirations (seen in mobile phone promotional videos since the 1990s) appear to be that the bots develop more autonomous behaviour, second guessing the needs of their users, and perform more complex reasoning when helping their user with a task, but even now, there is still minimal evidence of that.

Virtual Tutor

In the virtual tutor role, presented more fully in Chapter 11, the virtual human is supporting the learner over a lesson, course, or even a whole program or career. The virtual tutor should be able to move between the roles of teacher, coach and mentor, helping to direct and motivate the user, identify the gaps in the knowledge, how they best learn and where they struggle, and then deliver the learning and support the user needs to improve. All of this requires an effective memory of each student, better reasoning, a limited element of autonomous behaviour (although the user has final say), and, ideally, a more developed natural language ability, if a useful and trusting relationship is to develop. Whilst the simpler virtual tutors may still be within the virtual humanoid category, the more developed versions are beginning to justify the label of virtual human.

Virtual Non-Player Character

At first glance, the non-player characters (NPCs) within computer games appear to justify the virtual human label more than any of the systems considered above. They can look human, very human – or at least as human as the players themselves. Their behaviour, although designed to meet the goals of the game, is typically modelled closely on human behaviour. The increasing interest in so-called sandbox and open-world games with more open plots (e.g. Red Dead Redemption) requires the NPCs to be more behaviour- than script-driven, and to have their own goals and motivation. There is also an element of reasoning in how they choose their actions. However, although they are present as avatars within a virtual environment and have a degree of agency, there is always the game engine pulling the strings, and they have no real sense of embodiment or self-determination, let alone any inner life. However, if the sandbox game is pushed into the extreme of the true open world, then there is potential for such characteristics to possibly develop – as some of the science fiction examples considered in Chapter 2 show.

Virtual Life Coach

The Virtual Life Coach (VLC) represents a more developed form of the Virtual Tutor. It may include virtual tutor capabilities, or it may refer the user to a Virtual Tutor more knowledge on a particular subject, as required. With the Virtual Life Coach, there is a far greater emphasis on developing the trust and bonding between the VLC and the user. Current 'life coach' apps, such as Woebot and Wysa, represent a very basic form of VLC operating in a very limited knowledge domain. As VLCs develop, more sophisticated reasoning, autonomous behaviour (in terms of developing and suggesting courses of action, or gathering information), natural language and even emotional responsiveness could be expected. Whether the VLC ultimately merges with the Virtual Personal Assistant or they remain as two separate apps or persona is an interesting consideration.

Virtual Expert

The Virtual Expert is the logical progression of today's Virtual Assistants, Virtual Tutors, and customer service chatbots. The Virtual Expert is a virtual human which, whilst it is constrained to its domain of expertise, is more or less indistinguishable in knowledge, process, behaviour, look and sound from a physical human. Whilst lower grade customer service agent chatbots might still deal with the routine calls (there's no point in wasting resources), it is the Virtual Experts that might begin to displace the supervisors and middle managers. The potential impact of such virtual workers will be considered in more detail in Chapter 13. However, there is no suggestion that the Virtual

Expert requires any 'interior life', or any personality beyond that needed to create effective relationships with its customers, users and colleagues.

Virtual Persona

All the examples considered so far have been bots which represent some objective, best practice system, given the personality best suited to perform their task and working with the user. With the virtual persona (sometimes called a digital twin), the focus begins to shift to the personality being paramount, with possibly more subjective information and less-than-optimal performance and behaviours. A virtual persona could represent an extreme development of an NPC bot, a virtual actor even, or it could be developed from the very start to be a virtual persona. One possible application of virtual personas is to capture the knowledge and experience of those leaving the workplace – so that they are always available in a virtual version to those still working. Another application is to support dementia sufferers, either with the Persona of the sufferer or of a caregiver, but with the intention of leveraging the virtual persona's knowledge of the sufferer's life to help to keep memory pathways active. A third possibility is people using virtual personas of themselves to multiply their own work capability, cover for them when on holiday or in other meetings, or to start whole new lines of work. Here, traits such as the degree of autonomy, creativity, reasoning and motivation will be key. Figure 8.3 shows the virtual humans trait profile of a typical virtual persona – contrast this with that of the intelligent speaker shown at Figure 8.2.

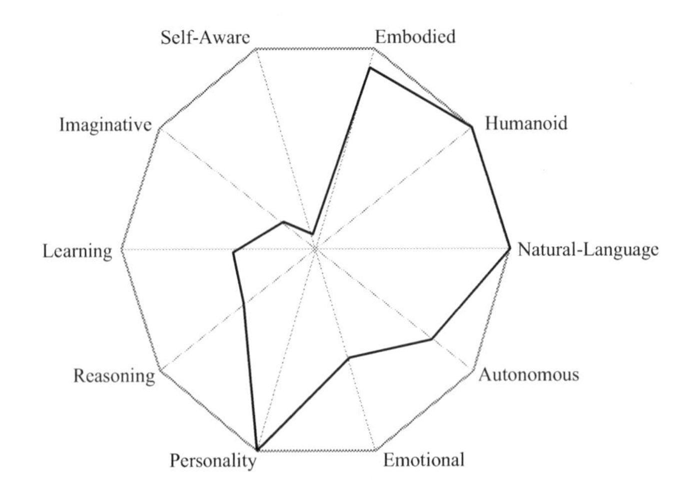

FIGURE 8.3 Virtual persona profile.

Virtual Person

Whilst the virtual persona represents a specific human being, and the Virtual Expert an expert in one particular field, the Virtual Person is a more general-purpose worker and personality. It represents a *tabula rasa*, a blank slate not inheriting a set of someone else's memories, or being bound to a single domain as is a Virtual Expert, but being able to learn or be taught in the same (or better) way as a physical human to do a task, and build its own memories as a result. As the needs in one activity fade, it can retrain for new tasks – and drop in and out of a robot body as the tasks require. With the arrival of such general purpose 'beings', the ethical and moral issues of virtual workers are likely to come to a head, and such ethical issues will be considered in more detail in Chapter 9.

Digital Immortal

The final category of virtual human is the Digital Immortal. A development of the virtual persona where the virtual human is, to all intents and purposes, indistinguishable from its physical progenitor. As such, and with the right hosting package, there is no reason why the virtual human, and the essence of any physical human it may be based on, should not become immortal. This will be considered in more detail in Chapter 12.

And Sentience?

What should be noted in all these examples is that whilst they represent a generally increasing level of sophistication, there seems to be no absolute requirement for them to become sentient. In terms of the AI landscape, the more developed forms are likely to pass Challenge 1, in terms of humanness, and even pass Challenge 2, in terms of being general purpose, but even without the latter, there is still probably a lot which can be achieved. By the later categories, an increasing degree of self-awareness would be expected, but whether this crosses over into true sentience is something which will be discussed in more detail in Chapter 13. But the lack of a requirement for sentience suggests that there is no particularly novel development required to move from category to category, at least to a useful level of performance, until the requirements of Artificial General Intelligence (also discussed in Chapter 13) can be achieved.

Transhuman Space Revisited

It is, perhaps, interesting to map the types of virtual human described above onto the typology presented in *Transhuman Space* (Pulver, 2002), as introduced in Chapter 2. Pulver describes a variety of AI types which map onto the different types of Virtual Humans discussed above. For his infomorphs ('living', that is, a sentient or near-sentient, software code), his Mind Emulations map onto the virtual personas and Digital Immortals. Pulver's types of Mind Emulation provide a useful level of further detail; the Ghosts are the perfect virtual personas (completely perfect copies), the Shadows are low-resolution versions (perhaps the current attempts at virtual personas are just barely reaching the Shadow stage). The Fragments are the failed copies (and just working with current virtual personas shows how easily such Fragments can be made when code is wrong or data is missing) and the result is a low-resolution, but distorted, version of the original source. However, Pulver predicates his Mind Emulations on the sort of Mind Uploading described in Chapter 13 – the Virtual Humans approach presented here has more in common with his eidolons.

Pulver's digital native AIs map onto Virtual Experts, Virtual Persons and the most advanced forms of Virtual Assistants and Tutors. Even the most developed forms of these currently achievable are, perhaps, not even at Pulver's Non-sapient Artificial Intelligence level (non-intelligent, unemotional, primarily, just a tool), and his SAI (Sapient Artificial Intelligence) level seems well out of reach for now. What is interesting is that his definition of Low-sapient Artificial Intelligence is intelligent but has difficulty with emotions. It would seem that a well-developed virtual human could readily both display and read human emotions and show required levels of empathy. Perhaps his Low-sapient Artificial Intelligence is better mapped to a virtual human with AGI capability, but short of sentience.

VIRTUAL HUMANS THROUGH DIFFERENT LENSES

This section considers how virtual humans appear when viewed through a number of different lenses typically applied to physical humans, and what this might reveal about how 'human' they really are. The lenses considered are:

- The lens of trust in technology;
- The species lens; and
- The lens of personhood.

The Lens of Trust in Technology

Lankton et al. (2015) investigated the levels of human trust in technology systems, and whether the degree of humanness matters in choosing how to trust such systems. In particular, they look at social presence theory (where a system is perceived as more sociable and warm if it gives the sense of another person being psychologically present), and social response theory (where people may treat and respond to a technology with higher social presence as though it were human). They define humanness simply as having 'the form or characteristics of humans' and show that technologies can differ in humanness and that users:

> [W]ill develop trust in the technology differently depending on whether they perceive it as more or less human-like, which will result in human-like trust having a stronger influence on outcomes for more human-like technologies and system-like trust having a stronger influence on outcomes for more system-like technologies (Lankton et al, 2015, 881).

Human-like trust is characterised by measures such as competence, benevolence and integrity, whereas system-like trust uses measures such as functionality, helpfulness and reliability. In each case, the desired outcome was for the user to see the system as useful (i.e., high in perceived value), enjoyable, trusting in intention (i.e., willing to depend on the system) and continuance intention (i.e., intent or willingness to use over the long-term).

The implications of this for virtual humans is a two-stage challenge:

- The need to demonstrate a social presence and elicit a social response, have social affordances (for example, speech), and affordances for sociality (for example, two-way conversations), if the user is to consider them high in 'humanness'; and then
- To be seen as more 'human' than 'system', the virtual human needs to focus on showing competence not functionality, not just helpfulness but benevolence (i.e., going beyond just helpfulness by having the user's best interests at heart), and integrity and not just reliability.

It should be noted that by stressing benevolence and integrity, instead of just helpfulness and reliability, a moral element is being introduced into the judgment, the system is being assessed in a more human way.

The Species Lens

Wilson and Nick Haslam (2013) have identified two distinct senses of humanness in their investigations into social psychological behaviour. Their interest stems from both understanding the divide between humans and animals, and in the changes in behaviour caused by different mental health conditions. The senses are a species-unique sense, called human uniqueness (HU) and a species-typical sense, referred to as human nature (HN).

> Human uniqueness, by definition, refers to those traits that distinguish humans from animals, exemplified by refinement, moral sensibility, self-control, and rationality. Uniquely human traits are judged as acquired through learning, as requiring maturity for their expression, not prevalent in the population, and culturally specific. In contrast, HN reflects the biologically-based human essence—the fundamental or essential attributes of the human species—some of which may be shared with animals. Traits judged as part of HN reflect emotional responsiveness, prosocial warmth, cognitive openness, and individuality. HN traits tend to be regarded as innate, prevalent within cultures, universal across cultures, and positive or socially desirable (Wilson and Nick Haslam, 2013, 373).

Essentially, HN is a measure of what separates humans (and animals) from robots, and HU is what separates humans (and potentially robots) from animals. HN is more 'animalistic', HU is more 'rational'. The two can be shown as a simple set of axes, as in Figure 8.4.

Wilson and Haslam identify 60 traits and behaviours which can be used to evaluate the levels of HU and HN based on how much they are seen more in humans, animals or robots. These range from High-HN behaviours, such as sleeping, eating, drinking, sex and relationships, to High-HU behaviours, such as performing complex and repetitive tasks, making things and following rules.

The development of virtual humans from a primitive virtual humanoid/chatbot-type system, through a mature virtual human and to a potential virtual sapiens could be seen as a journey across the

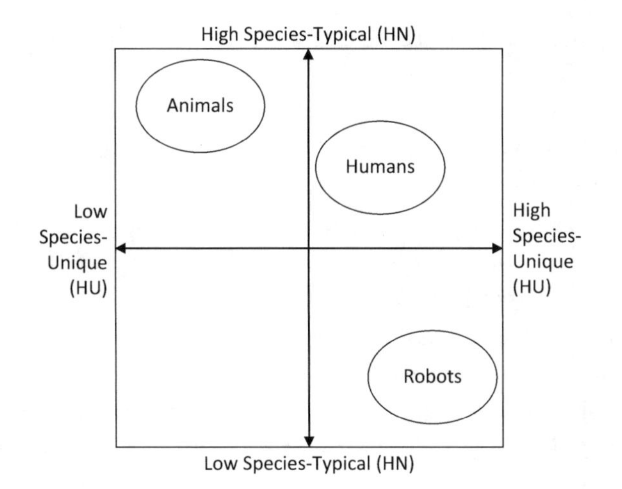

FIGURE 8.4 The HU-HN matrix.

landscape in Figure 8.4 from a current High HU/Low HN to a High HU/HN state, and, perhaps, even shedding some of the more extreme High HU characteristics. Looking at the traits that Wilson and Haslam present, this means focusing virtual human development on areas such as communicating, environment awareness, teaching, game playing, the creative arts and social relationships and helping others, without necessarily the need to move into more extreme HN behaviours, such as fighting, eating, drinking and sex.

The Lens of Personhood

A further useful lens through which to consider the virtual human is that of personhood. Harré (1997, 6) defines personhood as a threefold phenomenon, 'a sense of personal distinctiveness, a sense of personal continuity and a sense of personal autonomy'. Rutkin (2016) presents six attributes for animal personhood developed by Kristin Andrews: subjectivity, rationality, personality, relationships, autonomy and narrative self. Each of these will now be considered in turn.

As a virtual human grows its own memories (or inherits those of a human progenitor) it will show *subjectivity* and a unique perspective and point of view, accompanied by an emotional landscape and the ability to reflect that. Its decisions and observations will be influenced by its own motivations and world view.

Much of the focus of AI research has been in developing the ability to show *rationality*, thinking and logical reasoning; the challenge with virtual humans is to show those at a human level, and to know when not to behave too logically or rationally.

Whilst it is relatively trivial to program a distinctive *personality* and individual character in the virtual human, the more interesting goal is in seeing how this might emerge as the virtual human lives its life and learns its preferred way of behaving and achieving its goals.

In terms of *relationships*, there is already evidence of the bonds that form between physical humans, chatbots and robots (Ciechanowski et al., 2018). At present, these tend to be from the physical human to the virtual human, with the 'care' that a bot shows a human being a relatively programmed trait. Again, an area for future development would be to see how the virtual human relates to people as part of its own motivational and needs model, and effectively learning to care and bond with those whom it finds best contribute to its motivational needs.

Furthermore, the motivational model is again at the heart of the development of true *autonomy*. Whilst, as discussed in Chapter 1, a driverless car exhibits a high degree of autonomy within a specific domain, it has no ability to reject the trip to the supermarket and just go for a drive in the country instead. A true virtual human would be expected to have such a level of autonomy.

The *narrative self* has already been identified in Chapter 5 as a key element lacking in almost all current AI and virtual human research, and a potential key objective if true virtual humans are to be developed.

On this basis, it would seem that a good case could be made for allowing a well-developed virtual human the status of personhood. There will always be the challenge that the virtual human has 'just been programmed that way', and, ultimately, it may be that the virtual human arguing their own case for personhood may be the only accepted measure.

MORE THAN THE SUM OF ITS PARTS?

This challenge that a virtual human is 'just a program' is, perhaps, at the heart of the debate about whether a virtual human is 'truly' human, no matter how sophisticated it may be. A competent virtual human should readily pass the lens of trust without any need to debate its human nature (albeit accepting that it might show a high level of 'humanness'). The species lens is more challenging, but much of Chapter 4 was about showing how the more HN species-typical/animalistic characteristics could be reproduced in a virtual human. The true challenge is, indeed, in assessing not whether the virtual human just shows a high degree of 'humanness' but whether it shows enough subjectivity, personality, relationship building, autonomy and a sense of narrative self in order to be considered as its own person.

The following chapters will consider some of the more social aspects of virtual humans in more detail, especially Ethics (Chapter 9) and Identity and Agency (Chapter 10) before going on to look at two specific use cases in more detail (Education and Immortality) and then the possible future evolution of the virtual human.

REFERENCES

Ciechanowski, L., Przegalinska, A., Magnuski, M., & Gloor, P. (2018). In the shades of the uncanny valley: An experimental study of human–chatbot interaction. *Future Generation Computer Systems*. doi:10.1016/j.future.2018.01.055.

Harré, R. (1997). *The Singular Self: An Introduction to the Psychology of Personhood*. London, UK: Sage Publications.

Lankton, N. K., McKnight, D. H., & Tripp, J. (2015). Technology, humanness, and trust: Rethinking trust in technology. *Journal of the Association for Information Systems, 16*(10), 880.

Leviathan, Y. (2018). *Google Duplex: An AI System for Accomplishing Real-World Tasks Over the Phone*. Google AI Blog. Available online https://ai.googleblog.com/2018/05/duplex-ai-system-for-natural-conversation.html.

MacBlain, S. (2018). *Learning Theories for Early Years Practice*. London, UK: Sage Publications.

Miron, A. M., McFadden, S. H., Hermus, N. J., & Buelow, J. (2017). Contact and perspective taking improve humanness standards and perceptions of humanness of older adults and people with dementia: A cross-sectional survey study. *International Psychogeriatrics, 29*(10), 1701–1711.

Pulver, D. L. (2002). *Transhuman Space [Role-playing game]*. Austin, TX: Steve Jackson Games.

Rutkin, A. (2016). Almost human? *New Scientist, 231*(3080), 16–17.

Slater, S., & Burden, D. (2009). Emotionally responsive robotic avatars as characters in virtual worlds. In *Games and Virtual Worlds for Serious Applications VS-GAMES'09. Conferencein* (pp. 12–19). IEEE.

Wilson, S. G., & Nick Haslam, N. (2013). Humanness beliefs about behavior: An index and comparative human-nonhuman behavior judgments. *Behavior Research Methods, 45*(2), 372–382.

III

Issues and Futures

INTRODUCTION

Part III, Issues and Futures, is the final section of the book which examines considerations such as identity and ways of dealing with the complexities and ethical challenges of these liquid technologies. It begins by exploring some of the ethical, moral and social issues and dilemmas which virtual humans introduce in Chapter 9 'Digital Ethics'. In Chapter 10 'Identity and Agency', it examines perspectives on, and studies into, identity, as well as considering work on identity positions in virtual worlds.

The final three chapters of the book take a broader view, by exploring the possible impact that virtual humans could have on our everyday lives, and even on our relationships. Chapter 11 'Virtual Humans for Education' examines the probable impact of virtual humans on supporting learning. Building on the issues of relationships Chapter 12 'Digital Immortality' explores the emotional, social, financial and business impact of active digital immortality on relations, friends, colleagues and institutions. The final chapter, Chapter 13 'Futures and Possibilities' considers what the impact of virtual humans might be, and what significant other developments might take place, within three successive time-frames: 2019–2030; 2030–2050; and 2050–2100 and beyond. The chapter then examines how the three main challenges to the developments of virtual humans identified in Chapter 2 might be addressed, namely improving humanness, developing an Artificial General Intelligence and the emergence of Artificial Sentience and Virtual Sapiens.

CHAPTER 9

Digital Ethics

INTRODUCTION

This chapter will explore some of the ethical, moral and social issues and dilemmas which virtual humans introduce. There is currently very little understanding of exactly how virtual humans might be used in teaching, research, business or government organizations. It is also not clear how the use of virtual humans affects people's disclosure of information or the quality of that information. What is certain, however, is that discussions around ethics must be at the forefront of the use of virtual humans. This chapter examines what it means to undertake ethical research in the context of virtual humans and examines issues such as technical ethics, design ethics, legal issues, and social concerns, including honesty, plausibility and the nature of consent.

RECENT CONSIDERATIONS OF ETHICS AND VIRTUAL HUMANS

There has been increasing realization that there needs to be a delineation between different types of ethics: robotics ethics, machine morality and intimacy ethics are all new areas that being examined.

Robotics Ethics

Malle (2016) argues that robot ethics needs to examine ethical questions about how humans should design, deploy and treat robots.

He suggests that two issues need to be explored, both of which apply to virtual humans:

- Ethical questions about how humans should design, deploy and treat robots.
- Questions about what moral capacities a robot should have.

Robot ethics features such topics as ethical design and implementation as well as the considerations of robot rights. Hern (2017) reported that The European Parliament has urged the drafting of a set of regulations to govern the use and creation of robots and artificial intelligence. The areas that need to be addressed are suggested to be:

- The creation of a European agency for robotics and AI;
- A legal definition of 'smart autonomous robots', with a registration for the most advanced;
- An advisory code of conduct to guide the ethical design, production and use of robots;
- A new reporting structure for companies requiring them to report the contribution of robotics and AI to the economic results of a company for the purpose of taxation and social security contributions; and
- A mandatory insurance scheme for companies to cover damage caused by their robots.

The report also takes a special interest in the future of autonomous vehicles, such as self-driving cars, but as yet, there seems relatively little detail about how this might be implemented or developed or, indeed, what the relationship between AI and virtual humans is. However, in the autumn of 2017, Sophia, a humanoid robot, gave a speech at the United Nations, to prompt the recognition that there needs to be more debate, as well as legislation, in this area.

Machine Morality and Intimacy Ethics

Machine morality explores issues about what moral capacities a robot or virtual human should have and how these might be implemented.

It also includes issues such as moral agency justification for lethal military robots and the use of mathematical proofs for moral reasoning (Malle, 2016).

There are a range of debates on whether robots and virtual humans can have both emotions and empathy. For example, Prinz (2011) suggests that empathy is not needed for moral judgment, whereas moral judgment does require emotion. However, Docherty (2016) argues that robots and robot weapons lack both emotions and empathy and therefore cannot understand the value of human life, although should a robot (or virtual human) be judged to show emotions or empathy, then this may no longer hold.

Considering intimacy ethics engaging with virtual humans offers people opportunities to connect with something emotionally and feel supported, even loved, without the need to reciprocate. Some authors find relationships with virtual humans to be uncanny (Turkle, 2010) and, for many, people marrying robots and substituting loved ones seems a step too far. Although this kind of inter-relationship is becoming more common (Levy, 2008), it introduces challenging questions about what it means to be human. Malle (2016, 244) argues that:

> Any robot that collaborates with, supports, or cares for humans – in short, a social robot – poses serious ethical challenges to the human design and deployment of such robots, and one of the most important challenges is to create a level of moral competence in these robots that is adequate to the application at hand.

Borenstein and Arkin (2016) argue that main issues that need to be considered are affective attachment, responsiveness and the extent and duration of the proximity between the human and the robot. For example, Michel (2013) reported U.S. military personnel forming attachments with bomb disposal robots. Emotional connection has also been found to be one of the strongest determinants of a user's experience, triggering unconscious responses to a system, environment or interface. Thus, an ethical concern is the extent to which a robot or virtual human affects a human's autonomy, for example, whether those people who have trouble forming relationships will develop an over-dependent relationship with a virtual human or robot, which reduces their ability to act as an autonomous human being.

Technical and Design Ethics

Virtual humans are designed using human decision criteria and, therefore, ethical behaviour need to be 'designed-in' to virtual humans. However, in the context of the range of virtual humans that are being developed or considered, designing appropriate and effective ethics standards are complex and far reaching. For example, if autonomous combat robots or AI are deployed by the military, whose fault is it if they mistakenly attack or fail to distinguish correct targets?

Riek and Howard (2014) suggest that design considerations should include:

- Reasonable transparency in the programming of the systems.
- Predictability in their behaviour.
- Trustworthy system design principles across hardware and software design.
- Opt-out mechanisms (kill switches).

Legal Issues

Legal frameworks for research exist at local, national and international levels, which institutional review boards will be able to advise on. One particularly important framework, for example, is the European Union's (2014) ruling that search engines must respond to user requests for the removal of search results related to their names. This is otherwise known as the 'Right To Be Forgotten' ruling. It remains to be seen how this ruling and others like it might influence the dissemination and longevity of research findings. What is most important here is that the legal context is ever-changing, and, thus, research for a digital age, in some respects, requires researchers to be aware of current debates and plan research projects and data management accordingly.

The digital world, too, opens up an unimaginably diverse range of legal complexities. One example is a virtual human involved in social media or undertaking business transactions when the physical entity is deceased, which will be discussed in more detail in Chapter 11. Yet, currently, there is not a single global jurisdiction, and the laws that apply in one country or state may not necessarily apply elsewhere, which means that the use and abuse of virtual humans across the world, particularly in terms of those who create virtual humans for illegal purposes for use in a different country from their own, are unlikely to be prosecuted. Furthermore, it is not clear if the actions of the virtual human may result in criminal liability. This raises the question again about the driving force behind the virtual presence; the dead can generally have no criminal liability, but others acting behind the shield of a digital presence may have. Savin-Baden et al. (2017, 23) ask:

> What if a claimant were to attempt an action in libel against a virtual presence? In UK law defamation is a civil offence, so the claimant would be seeking to sue an entity purporting to be an individual that no longer exists. How might this be possible? If the entity was acting under the deceased's autonomous control, for example a perpetuating social media presence, then the claimant may attempt to sue the dead person's estate ... What liability do website operators have for defamatory comments they host? What if the potential action sits outside the jurisdiction of the English court, for example elsewhere in the EU, or in the US? The answers to these questions must be framed within the context of a range of caveats and conditions.

Riek and Howard (2014) suggest that decision paths in the creation of robots and virtual humans must be able to be reassembled for the purposes of litigation and dispute resolution. However, as well as legal issues, there are also ethical concerns that relate to which stand point we adopt.

ETHICAL STANDPOINTS AND VIRTUAL HUMANS

This section outlines approaches to ethics and links them to issues connected to virtual human development and research. These standpoints are outlined below; however, the work of Ess (2017a, 2017b) in digital media ethics and his work on phronesis (Ess, 2016) is particularly useful in guiding thinking about ethics and virtual humans. Öhman and Floridi (2018) have also written an article which suggests they are providing an ethical framework for the digital afterlife industry, which offers some interesting perspectives, but little in terms of an actual ethical framework. The challenge of understanding ethics in the context of virtual humans is not just in the issues of design, decision-making and responsibility, but the ethical standpoint you adopt.

Utilitarian

Utilitarianism is an ethical theory where the focus is on the outcomes and, thereby, assumes ethics can be governed by deciding what is right and wrong. This stance argues that the most ethical choice will be that which produces the greatest good for the greatest number of people.

Deontological

Deontology focuses on the rightness or wrongness of actions themselves, rather than the consequences, as in as utilitarian ethics. What counts as being right and wrong is guided by norms or duties, but sometimes the duties are not logical.

Virtue Ethics

Virtue ethics is a broad term for theories that emphasise the role of character and virtue rather than focusing on consequences or actions. The idea is that the focus is on someone displaying virtues and being humanitarian in a variety of contexts over time. However, the ability to cultivate the right virtues is affected by education, society, friends and family.

Situational Ethics

The focus here in situational ethics is on the situation or context of an act when evaluating it ethically, rather than judging it by moral standards. Thus, decision-making should be based upon the circumstances of a particular situation, and not upon law, but upon love. Situational ethics is based upon 'God is Love' in 1 John 4:8 in the Christian bible. The idea here is that people and context should be based on the situation and loving someone despite their actions, rather than focusing on the legal or moral stance.

Discourse Ethics

Discourse ethics refers to the ideal conditions in which discourse can occur and therefore centre around the agreements reached through the exercise of such discourse. Thus, ethics and ethical norms are created through discourse.

Table 9.1 compares different types of virtual humans with the differing ethical approaches.

TABLE 9.1 Types of Virtual Humans and Approaches to Ethics

	Utilitarian Ethics	**Deontological Ethics**	**Virtue Ethics**	**Situational Ethics**	**Discourse Ethics**
Simple Chatbot	If this benefits many people, then it is acceptable to use it.	As long as its actions are innocuous, moral and of benefit, this could be used.	The person programming this must a virtuous person.	If this benefits people in a given context, then it is acceptable to it.	It is unlikely that a chatbot will be able to create discourse.
Conversational agents	This may benefit many people, but it will depending on what is being discussed, and the consequences of that discussion.	This may be of benefit as an information giving agent but not if the conversation involves anything moral.	It is unlikely that the agent can make virtuous conversation on sensitive topics, and this could be a risk.	If the context is appropriate, this can be used.	This may be able to create discourse, but it is unlikely it would be able to deal with ethical issues.
Autonomous agent	Too much autonomy could be dangerous, as it is difficult to predict the consequences of an agent.	An autonomous agent would be too dangerous, as it does not have a moral code or sense of duty.	Virtues are affected by education, class and context and, therefore, it would be difficult to devise a virtuous agent.	Given an autonomous agent is unlikely to be able to understand context, this may pose a risk.	This may be able to create discourse, but ethical issues discussed could be risky.
Pedagogical agent	It is not clear if using an agent for teaching will have the right consequences, i.e., effective learning for most people.	Using this would be acceptable if the agent was programmed to follow actions clearly.	Teachers differ in the virtues they dispel, therefore, different types of virtuous agent would be needed.	This may work more effectively in some teaching contexts, but not others.	This could be used to help students understand the difficulties of discourse ethics and agents.

PRACTICAL IMPLICATIONS

Experimentation and exploration of the use of virtual humans has been explored for many years, but the ethical concerns in this area remain troublesome. Indeed, as Riek and Howard (2014) note, in the U.S., consumer robots developed by the industry require little ethical intervention before being sold. Yet there are practical concerns that still need to be addressed.

- When using virtual humans for mentoring or therapeutic reasons, such as relieving stress or improving personal skills, it is vital to undertake a risk assessment to ensure the possible impact on people when the virtual human is longer available to them.
- If virtual humans are being used as physically assistive robots for people with disabilities, it is important to consider the privacy of the client. For example, robots assisting with dressing and bathing should ideally not have video monitors or cameras, but such systems are probably essential in order to provide the sensory data for them to carry out their task. Further, the issue of emotional attachment to the virtual human needs to be considered for those who are living alone and may be lonely.

The virtual human needs to be designed in such a way as to reflect gender, race and ethnicity. To date, this issue has been rarely addressed and there has been much discussion by users and researchers in virtual worlds, such as Second Life, about the Hollywood-driven forms of stereotyping, for example, Waddell et al. (2014). It is vital that this issue is addressed and, where possible, that users can choose the gender, race and ethnicity of the virtual humans they engage with.

ETHICAL GUIDELINES AND RESEARCH

As described in Chapter 8, virtual humans can be used in a wide variety of ways, such as virtual patients, non-player characters, virtual survey takers, virtual mentors/therapists. However, there are relatively few guidelines for undertaking ethical research in relation to virtual humans.

Whilst approval from ethics committees and review boards is satisfactory at one level, some of the deeper ethical concerns are not addressed, such as how consent can really be 'informed' and issues, such as who is responsible when harm or injustice occurs. A useful starting point is The Association for Internet Researchers. They have developed a list of 63 questions specific to online research that digital researchers should ask themselves before beginning any research project (Markham and Buchanan, 2012, 8–11). In terms of this guidance, the following questions should be considered in terms of research using virtual humans:

- How are virtual humans being accessed, for example, on open or private networks?
- How are participants situated in relation to the virtual human?
- What is the relationship between the research and participants?
- What kinds of interaction will participants have with the virtual humans?
- Who will be responsible if harm or offence is caused by the virtual human?
- What kind of privacy and consent will be needed for participants engaging with virtual humans?
- How will participants' overreliance or over-engagement with the virtual humans be managed?
- To what extent is a virtual human considered to be a person?
- To what extent is digital data an extension of personhood?
- To what extent are virtual humans classed as human subjects?
- To whom does data collected by a virtual human in a virtual world belong?

These questions need to be considered in the context of each virtual human creation and research study and also in relation to differing approaches to ethics.

Ethical Behaviour

This is defined by people and contexts, and whilst posting messages on social media can be seen to be one's own business at one level, at another, it can be offensive, cause offence and result in others taking action if what is posted is deemed to be unethical. There are two major issues: errors of commission and errors of omission. *Errors of commission* include practices, such as using anonymous sources, phone hacking and making up sources.

Errors of omission include not asking tough questions of politicians and failing to exclude unnecessary violence from films and news clips. This appears to offer a straightforward distinction, yet, in fact, digital ethics is more complex than this.

Ethical Practice

This means respecting the needs of the individuals involved in research – whether participant or researcher – yet 'participants' and, indeed, 'researchers' are not homogenous groups but rather individuals with different beliefs about their needs in research. Therefore, there is a need for the ethical dilemmas associated with digital research to be made explicit, and for the challenging of ethical decisions in research to not only be accepted, but commonplace. It is only in doing so and, thus, opening oneself up to new points of view, that truly ethical work can be undertaken. At its most fundamental level, ethical research that includes human subjects is considered to be research that acknowledges the fundamental right of human dignity, autonomy, protection, safety, maximization of research benefits and minimization of harms (United Nations, 1948; National Institute of Health, 1949; World Medical Association, 1964; National Commission for the Protection of Human Subjects of Biomedical and Behavioural Research, 1978). Potential harms are taken to include physical harms (e.g., pain, injury), psychological harms (e.g., depression, guilt, embarrassment, loss of self-esteem), or social and economic harms (e.g., loss of employment or criminal charges) (Savin-Baden and Major, 2013). These rights have traditionally been assured through the following procedures, but, in fact, are troublesome in the face of virtual human research.

Anonymity

Anonymity means that no identifiable data (e.g., name, address) is collected as part of the research study, and therefore no identifiable data can be shared.

Confidentiality

Confidentiality involves the collection of identifiable data about research participants, which is then restricted to particular individuals depending upon the confidentiality agreement.

Informed Consent

Informed consent involves individuals consenting to participate in research when fully informed of the processes, benefits and harms involved.

Inherent in the principle of informed consent is the understanding that the participant can revoke consent, usually up until the study is completed and findings are published. In terms of virtual humans' research, as with all research, it means consent is ongoing and does not end when participants sign an informed consent form; rather, it is a process of continual negotiation. It may be achieved by full disclosure, adequate compensation and voluntary choice but, in practice, it means ensuring a balance between over informing and under informing. Human informed consent when engaging with virtual humans needs be a clear fit with purpose of the research and design objectives.

Minimal Risk

Minimal risk means that the risk associated with the study is similar to that typically encountered in everyday life. In terms of virtual humans' research, it is important that people are not deceived in unethical ways that may cause harm.

Honesty

Honesty, as a concept, allows us to acknowledge not only the cyclical nature of 'truths', but also that the nature of honesties is defined by people and contexts. It also helps us to avoid the prejudice for similarity and against difference in data interpretation. Furthermore, data about ethics, conduct and accountability can be distinguished by differences of theory and practical action, but they can never actually be isolated from one another. Much virtual human research demands that we engage with deceptions – our own and those involved in the research – and this in turn forces us to consider how we deal with such (benevolent) deception. It is also vital that researchers are aware of a user's ability to project traits on to a virtual human, such as that it cares about them, and, therefore, honesty also includes the need to avoid anything that may constitute deception.

Privacy

Privacy, as it refers to acceptable practices with regards to accessing and disclosing personal and sensitive information, has become a space of debates in the digital age. Often, data in the public domain is considered to be freely available for use for research purposes, without informed consent, so long as the individual providing the data is protected from harm as far as possible. The term 'found data' is useful here. Found data (Savin-Baden and Tombs, 2017) is defined as data available in the public domain; observations of people in public spaces such as stations, supermarkets and flash-mobs. These data are often found through covert-based research methods. Thus, there are ethical difficulties inherent in using them. The use of 'found data' is often justified based upon the transparency-and-choice concept, or the belief that the right to privacy is actually the right to control which information is released; in essence, the belief that if an individual puts data out in the public domain, they relinquish control of that data. Data perceived to be 'public' has been used for research purposes based on the justification that it does not include human subjects but rather involves the study of secondary data already available in the public domain. In virtual human research, it is vital to be clear how data are found, located, analysed and presented.

Plausibility

Plausibility is also of concern in virtual humans' research – for example, whether it is known who is being interviewed in an online setting and if this matters or not. For example, if participants are unidentifiable, it might be seen that the research lacks rigor because neither the population nor the demographics are known. The question here is, is the research viable, realistic, possible? Is it plausible to undertake it, to do so honestly and publish the findings in a peer reviewed journal?

Research Governance

Research governance is the process by which organizations, businesses and universities ensure all research activity within the organization complies with all relevant legislation. It is designed to supervise and administrate the way that research is managed so that participants and researchers are protected, and accountability is guaranteed as far as possible. Research governance requires:

- The provision of pre- and post-study information sheets (Information for Participants forms and Consent forms). These will give details of the research, the researcher and the implications for all participants. Where English is not the first language of participants, an interpreter will be used on behalf of the participants;
- The confidentiality of information supplied by research subjects and the anonymity of respondents will be respected;
- The research participants' involvement will be seen as participating in a voluntary way, free from coercion;

- Avoidance of harm to research participants;
- Ensuring adherence to ethical procedures for research misconduct, complaints or appeals;
- Conflicts of interest being made explicit; and
- Research participants being informed about any possible risks.

Much of the organization of research governance is about ensuring the prevention of harm, but, as yet, there is little included in research governance guidelines that considers the implications of using virtual humans, or, ultimately, of the potential harm to the virtual human or the robot.

Disclosure

Student willingness to disclose sensitive information to virtual humans has been attributed partially to virtual humans being *almost* like a person (Savin-Baden et al., 2013). Culley and Madhavan (2013, 578) have cautioned that:

> [A]s the agent becomes increasingly morphologically similar to a human, it is also likely that operators will engage in correspondence bias more frequently by ascribing human motivations, reasoning abilities and capabilities to this non-human system.

Corritore et al. (2003) suggest that the concepts of risk, vulnerability, expectation, confidence and exploitation play a key role in information disclosure in an online environment. A study by Savin-Baden et al. (2015) used a 2D website with a variety of avatar choices to explore student responses to pedagogical agent-mediated surveys on sensitive topics (finances, alcohol, plagiarism, drugs and sexual health) over varying periods of time. Either the survey was delivered in one sitting (short-term sitting), or three questions were delivered every three days over a period of two weeks (long-term sitting). Findings revealed that participants disclosed more information on the most sensitive topics when engaging with the pedagogical agent over the longer time period, emphasising a clear need for further study to be undertaken that examines the long-term impact.

THE 'HUMAN SUBJECT' AND VIRTUAL HUMANS

Early ethical guidelines developed the idea of the 'human subject' as a response to injurious medical experiments on humans. However, there are diverse arguments about the role, position and interaction of a virtual human in digital settings. There are also queries about whether informed consent is needed when engaging with virtual humans because the virtual human is not seen as having human values and interactions. The question to consider is perhaps whether the virtual human is seen as a stand-in for the researcher, in which case, it is likely that informed consent needs to be sought. Further, will informed consent need to involve informing participants whether they are being interviewed by a human or by technology? And might the use of virtual humans increase or minimize the potential harm to the participant?

It is clear that the use of virtual humans is resulting in complex questions about the nature of personhood. However, these questions and dilemmas also introduce questions about how data are collected and how they are represented and portrayed in research findings. For example, *representation* of data tends to refer to the way in which a researcher provides warranted accounts of data collected. Thus, the main way the term representation is used is in the sense of a proxy, the researcher is (re) presenting the views of the participants. This is often seen or presented by the researcher as being unproblematic. Whereas *portrayal* invariably is seen as the means by which the researcher has chosen to position people and their perspectives. Portrayal tends to be imbued with a sense of not only positioning but also a contextual painting of a person in a particular way. The use of virtual humans in research tends to focus more on representation than portrayal but there is a need to examine how exactly we portray interaction between participants and virtual humans. Recent research and literature has begun to explore some of the issues around personhood and representation.

Hasler et al. (2013) undertook a study in Second Life and compared interviews with an unintelligent pedagogical agent compared to an avatar controlled by a human interviewer. These findings indicated that participants were less willing to participate in pedagogical agent-facilitated interviews than human-facilitated interviews, but also that both approaches were equally successful in gathering information on participants' backgrounds. The suggestion from this study was the use of pedagogical agent interviewers could reduce workloads. However, there is presently very little understanding of exactly how pedagogical agents might be used ethically and effectively in research settings. What is certain, however, is that discussions around ethics must be at the forefront as artificially intelligent pedagogical agents continue to become 'researchers'.

CONCLUSION

The sphere of digital ethics, with or without the complication of virtual humans, remains complex, and it is as yet unclear what the impact might be on people's lives. What is clear is that there is little legal guidance, and, in many instances, current ethical guidelines supplied by governments, universities and professional bodies fail to deal with the fast-changing nature of digital ethics and the emergence of virtual humans. Perhaps being critical and asking questions are the most important concerns when considering ethical issues in relation to virtual humans. Central to criticality is the issue of agency, and it is this that is explored in Chapter 10.

REFERENCES

Borenstein, J., & Arkin, R. C. (2016). Robots, ethics, and intimacy: The need for scientific research. In *Conference of the International Association for Computing and Philosophy* (*IACAP 2016*), Ferrara, Italy, June.

Corritore, C. L., Kracher, B., & Wiedenbeck, S. (2003). On-line trust: Concepts, evolving themes, a model. *International Journal of Human-Computer Studies, 58*(6), 737–758.

Culley, K. E., & Madhavan, P. (2013). A note of caution regarding anthropomorphism in HCI agents. *Computers in Human Behavior, 29*(3), 577–579.

Docherty, B. (2016). Losing control: The dangers of killer robots. *The Conversation*, June 16.

Ess, C. M. (2016). Phronesis for machine ethics? Can robots perform ethical judgments? In J. Seibt, M. Nørskov, & S. S. Andersen (Eds.), *What Social Robots Can and Should Do*, Frontiers in Artificial Intelligence and Applications (pp. 386–389). Amsterdam, the Netherlands: IOS Press.

Ess, C. M. (2017a). Communication and technology. *Annals of the International Communication Association, 41*(3–4), s209–s212.

Ess, C. M. (2017b). Digital media ethics. *Oxford Research Encyclopedia of Communication*. doi:10.1093/acrefore/9780190228613.013.508.

European Union Directive 95/46/EC of 13th May 2014 on personal data – the protection of individuals with regard to the processing of such data.

Hasler, B. S., Tuchman, P., & Friedman, D. (2013). Virtual research assistants: Replacing human interviewers by automated avatars in virtual worlds. *Computers in Human Behavior, 29*(4), 1608–1616.

Hern, A. (2017). Give robots 'personhood' status, EU committee argues. *The Guardian*, January 17.

Levy, D. (2008). *Love and Sex with Robots: The Evolution of Human-Robot Relationships*. New York: Harper Perennial.

Malle, B. F. (2016). Integrating robot ethics and machine morality: The study and design of moral competence in robots. *Ethics Information Technology, 18*, 243–256.

Markham, A., & Buchanan, E. (2012). *Ethical Decision-Making and Internet Research: Recommendations for the AOIR Ethics Working Committee Version 2.0*, Association of Internet Researchers. Available online http://aoir.org/reports/ethics2.pdf.

Michel, A. H. (2013). Interview: The professor of robot love. Center for the Study of the Drone, October 5. Available online http://dronecenter.bard.edu/interview-professor-robot-love/.

National Commission for the Protection of Human Subjects of Biomedical and Behavioral Research (1978). Belmont report: Ethical principles and guidelines for the protection of human subjects of research. Available online http://www.fda.gov/ohrms/dockets/ac/05/briefing/2005-4178b_09_02_Belmont%20Report.pdf (January 18, 2016).

National Institute of Health (1949). *Nuremberg Code*. Available online https://history.nih.gov/research/downloads/nuremberg.pdf.

Öhman, C., & Floridi, L. (2018). An ethical framework for the digital afterlife industry. *Nature Human Behaviour, 2*, 318–320.

Prinz, J. J. (2011). Is empathy necessary for morality? In A. Coplan & P. Goldie (Eds.), *Empathy: Philosophical and Psychological Perspectives* (pp. 211–229). Oxford, UK: Oxford University Press.

Riek, L. D., & Howard, D. (2014). A code of ethics for the human-robot interaction profession (April 4). *Proceedings of We Robot, 2014*. Available online https://ssrn.com/abstract=2757805.

Savin-Baden, M., Burden, D., & Taylor, H. (2017). The ethics and impact of digital immortality. *Knowledge Cultures, 5*(2), 11–29.

Savin-Baden, M., & Major, C. (2013). *Qualitative Research: The Essential Guide to Theory and Practice*. London, UK: Routledge.

Savin-Baden, M., & Tombs, G. (2017). *Research Methods for Education in the Digital Age*. London, UK: Bloomsbury.

Savin-Baden, M., Tombs, G. & Bhakta, R. (2015). Beyond robotic wastelands of time: Abandoned pedagogical agents and *new* pedalled pedagogies. *E-Learning and Digital Media, 12*(3–4), 295–314.

Savin-Baden, M., Tombs, G., Burden, D., & Wood, C. (2013). It's almost like talking to a person. *International Journal of Mobile and Blended Learning, 5*(2), 78–93.

Turkle, S. (2010). In good company? On the threshold of robotic companions. In Y. Wilks (Ed.), *Close Engagements with Artificial Companions: Key Social, Psychological, Ethical and Design Issues* (pp. 3–10). Amsterdam, the Netherlands: John Benjamins Publishing Company.

United Nations. (1948). Universal declaration of human rights. Available online http://www.un.org/en/universal-declaration-human-rights/.

Waddell, T. F., Ivory, J. D., Conde, R., Long, C., & McDonnell. R. (2014). White man's virtual world: A systematic content analysis of gender and race in massively multiplayer online games. *Journal for Virtual Worlds Research, 7*(2). Available online https://jvwr-ojs-utexas.tdl.org/jvwr/index.php/jvwr/article/view/7096.

World Medical Association. (1964). *Declaration of Helsinki*. Helsinki, Finland: World Medical Association.

CHAPTER 10

Identity and Agency

INTRODUCTION

This chapter will explore the notion of virtual humans from a broad perspective of technology and identity by exploring perspectives on, and studies into, virtual reality and immersion, as well as studies about identity positions in virtual worlds. This chapter will also build on the work of Savin-Baden (2015); who has defined different types of identities, namely, Spatial identities, Networked identities, Bridged identities and Discarded identities. The relationship between virtual humans and physical humans is complex, with a sense of both overlaps and overlays. This chapter will argue that what is needed is not a static view of identity and agency but a liquid view; that they are multiple and fluid and overlap and shift according to context.

RETHINKING IDENTITY

Ideas and perspectives about identity change from decade to decade. There have been shifts away from the idea of a stable self towards a sense that identities change and move over time. The work of Bernstein (1996) was influential in arguing for the importance of pedagogic identities and the ways in which they are influenced within academe. Pedagogic identities are those that emerge from technological change and arise in contemporary culture and are therefore characterized by the emphases of the time. For example, in the traditional disciplines of the 1960s, students were inducted into the particular pedagogical customs of those disciplines, whereas pedagogic identities of the 1990s were characterised by a common set of market-related, transferable skills. In the 2000s, pedagogic identity would seem to be less strong and instead students define themselves through social media characterised by constant interaction and affirmation. Furlong and Davies (2012) argue that young people's engagement with new technologies is fundamentally bound up in their own identity. Identity creation and exploration is not only evident through representations on social networking sites but also the ways in which people accessorize themselves technologically. Haraway (1985) and Hayles (1999, 2012) have been at the forefront of discussions about identity in digital spaces, and Ito et al. have been influential in the work that examined how youth culture and identity might be understood. Thus, along with the raft of sociologists who have examined identity, there is a broad literature on identity, but relatively little that has analysed identities in relation to virtual humans. However, there has been work undertaken on identity in virtual worlds. For example, Steils (2013) developed a typology of learner identity in virtual worlds through showing how students seek to manage their identity through their avatar. The typology portrays learner identity across five dimensions and is summarised below:

Dimension One: Dislocated Avatars

The first dimension concerns the utilization of default avatars. In this dimension, the avatar was positioned as merely functional. No emotional attachment to the avatar was indicated; participants basically ignored the appearance of the avatar.

Dimension Two: Representative Avatars

The second dimension considers an understanding of the avatar as a representation of oneself in functional terms; although the avatar's appearance seemed closely connected to (or at least an approximation) of the physical world appearance of the user, the students described the relationship to the avatar in terms of functional representation and rendered the avatar as an object or tool.

Dimension Three: Avatars as Toys and Tools

The third dimension portrays an understanding of the avatar as a tool and object for playful engagement, as well as a status object. The avatar was, here, positioned as an object that could be customized and played with to take on varied appearances.

Dimension Four: Avatars as Extensions of Self

The fourth dimension regards students who declared their respective avatars as an extension of themselves, both visually, as well as emotionally. In their narratives, the avatar was described as closely related to the user not only in corporeal appearance and, possibly, by name, but the students also mentioned being emotionally and psychologically attached to their avatars.

Dimension Five: Avatars as Identity Extensions

In this dimension, students engaged with their avatars and the virtual world in terms of 'laboratories for the construction of identities' (Turkle, 1996, 184, 1999). Thus, the avatar in this dimension ventured into dimensions of exploring notions of potential, new or alternative identities for oneself both in the context of the virtual world or as a testing ground for the physical world, as well as exploring possibilities beyond its boundaries.

(Steils, 2013, 242–250)

These typologies could be adapted for use with virtual humans, but as yet this project has not been undertaken. However, a tentative adaption can be made, with two possible perspectives taken, that of the human user to the virtual human, and that of the virtual human itself:

- Dislocated Avatars: For the human, they see the avatar of the VH as simply a visual signifier and think nothing particular about its dress or skin. This is particularly the case where the avatar is deliberately robotic or takes a non-human form. For the virtual human, it is the case where the virtual human has no control over the avatar choice – that has been made by the programmer or user.
- Representative avatars: The human sees the avatar dressed appropriate for its role and interacts with it on the basis of its role, such as one dressed as a Neolithic woman within a simulation of a historic site. The virtual human is able to select an avatar which matches its role itself, or one suited to its current mood or task.
- Avatars as toys and tools: If the virtual human is able to change its avatar on a rapid basis to meet the whims of the user, especially when engaged in game playing or even sexual roleplaying, then the 'avatar as toy' dimension will be more dominant. For the virtual human, a relatively high level of mental ability is required for it to

recognize that a situation calls for 'play', and that it can change its appearance not in order to better fulfil a specific role but purely to increase the happiness and enjoyment of itself and of the humans it is interacting with.

- Avatars as extensions of self: At this level, the human is recognizing that behind the avatar is a virtual human with some sense of personhood – and that the virtual human's avatar is a manifestation of them. Likewise, for the virtual human, it suggests an even more developed sense of 'self' and the idea that some of its avatars represent a 'truer' sense of its self than others.
- Avatars as identity extensions: In some ways, a combination of avatar as toy and avatar as self – with a far more purposeful changing of identities in order to learn more about oneself. For the human, it is when they see the virtual human as co-explorer, mutually trying out new identities, and for the virtual human it is when they singly and deliberately take on new identities in order to understand and learn how humans respond differently to these new identities, and how these identities make they themselves feel and what they learn from their personal experiences.

FORMATIONS AND FORMULATIONS OF IDENTITY

What is increasingly evident is that identities change and shift according different to media and online contexts. There is no longer the sense that we have fixed identities. Instead, our identities shift and move with contexts, temporalities and the (virtual) worlds in which we live and work. Thus, these might be delineated as multiple and multifaceted, but perhaps shaped as follows.

Spatial Identities

Spatial identities are enacted through digital media, which tends to prompt a different kind of performance, invariably guided by the norms, cultures and affordances of both the software and the users of those spaces. There are the many mobile and polyvalent identities that are played in diverse media spaces, whether Twitter, Facebook, blogs or email. Largely, there will be some degree of crossover between them, and some kind of stability, even though we might choose to represent ourselves differently on Facebook compared with a work email conversation.

In terms of virtual humans, the context in which they are placed, such as learning or work, as well as the role for which they are used – teacher, or life coach, for example – will affect the way in which they are perceived and adopted by users. As with a physical human, it is also possible for a virtual human to have mobile and polyvalent identities across a variety of social virtual worlds and social media spaces, and it may even, in a human-like way, deliberately edit or represent itself differently on different platforms. Perhaps that degree of difference is even a (crude) measure of its sophistication as it recognizes the need or efficacy of having identities tuned to particular spaces and needs.

Networked Identities

Networked identities overlap across online and off-line networks, across school, work and spare time. Ryberg and Larsen (2008) have suggested that it is through such networks that individual identities exist and become real. Networked identities differ from spatial identities as they are specifically located in, and in relation to, given networks, rather than physical places or particular media.

Virtual humans would also be seen to be located within networks, such as schools, work or recreation. Whilst early virtual humans are tending to be created to operate just in a specific network (e.g., a game NPC does not appear on your Facebook feed) as they become more capable and more general purposed, such as the Virtual Person and virtual persona described in Chapter 8, then it is more likely that they will be operating in multiple networks and so develop more networked identities. How such virtual humans will manage and assimilate such networked identities is, again, an interesting aspect of their maturity.

Bridged Identities

Bridged identities create a link with other exterior worlds, such as virtual worlds, discussion forums and gaming worlds. With bridged identities, the bodily markers that are used to present ourselves in life, clothes, ethnicity, gender and speech may be represented (differently) in virtual spaces, but they also indicate choices about how we wish to be seen or the ways in which we might like to feel differently. However, as Nakamura (2010, 430) suggests, we might be aware that these kinds of media (games and virtual embodiment) create social factions of race and gender, while accurate images of gender and cultural realties might be rare.

A virtual human should have the same opportunity as a physical human to develop a bridged identity as it enters different virtual and social worlds, making the same choices (and, ideally, for the same reasons). However, it may be even more motivated in terms of identity extension than a physical human since it may have no 'root' culture, ethnicity or gender on which to base its virtual avatar or identity.

In many ways, the creation of a virtual persona could be considered a special case of a bridged identity. A virtual persona is a virtual human system which contains relevant memories, knowledge, processes and personality traits of a real person, and enables a user to interact with that virtual persona as though they were engaging with the real person. The virtual persona is developed to model the individual's personality traits, knowledge and experience that they have gained or have heard of, and also incorporate the individual's subjective and possibly flawed view of reality. The information provided is by the individual in a curated form and, as such, only data the individual chooses to share has been included.

Discarded or Left-Behind Identities

Žižek (1999), in his deconstruction of *The Matrix,* suggests the possibility that the deletion of our digital identities could turn us into 'non-persons' – but, perhaps, a more accurate idea would be one of becoming changelings, or having left-behind identities, rather than being seen as deletions. Thus, as we shift and move identities across online contexts, rather than deleting those identities that become superfluous, we tend to leave them behind. Such identities then become part of the junk spaces of the Internet, left behind and forgotten avatars, discarded blogs or Facebook profiles, along with the ones that remain when someone dies, which may bring memories, sadness or feel sinister for some.

As will be discussed in Chapter 12, these discarded and left-behind identities may become the foundations for Digital Immortals, and what in Chapter 2 Newton refers to as Thanograms, and Pulver as Shadows. It is also the case that the virtual human will be creating its own discarded and left-behind identities as it uses virtual worlds and social media spaces in the same way as physical humans. Indeed, some of the authors' past virtual human experiments have included virtual humans that tweeted and even blogged; these have not been explicitly deleted and so are already representing the digital traces and left-behind identities of early virtual humans.

Cyberspace and Virtual Humans

There still remain many debates about the nature on identities in cyberspace, with a wide diversity of opinion. For example, some scholars suggest that it is important to have a clear conceptual understanding of who we are in cyberspace, since without it, we risk being confused (e.g., Floridi, 2011). Those, such as Floridi, who argue for such a stance, seem to suggest that by separating and being honest about identities brings with it some kind of honesty or morality, yet this would seem misplaced.

We suggest understanding identities in relation to virtual humans as described above is complex and we need to have a new set of s/pl/ace identities which reflect squashed polarities, chaotic overlaps and new configurations of space and place, as well as an appreciation of virtual reality and immersions.

UNDERSTANDING AGENCY

Agency, according to Giddens (1984, 14), is the 'capacity to make a difference'. Giddens's theory of structuration is a social theory which suggests that the creation and reproduction of social systems is based on both structure and agents. Therefore, agency is inseparable from structure: agency is shaped by structure, while structure is produced, and reproduced, by human actions. Giddens identifies three dimensions of structure, namely, signification, domination and legitimation and three dimensions of interaction, namely, communication, power and sanctions, with, which the structural dimensions, are linked through *modalities* of, respectively, interpretive schemes and facilities. What is central to the structure-agency argument is that individuals and structures are inextricably linked. Take, for example, the use of devices. The electronic devices people use reflect the influence of social structures that are reproduced by individuals' conformance with accepted practice. Young people, for example, will see their phones as both accessories and necessities. Similarly, in terms of clothes, people working in offices will usually wear formal clothes, doctors in hospitals will wear white coats. These *structures of signification* inform our understanding of someone's role. However, devices and clothes do not simply indicate who a person is, but also convey important messages about *structures of domination*: the powers they are considered perceived to hold. Thus, a doctor's white coat will enable them to gain access to an intensive care unit and, in a military context, uniforms provide information about how one is expected to behave towards someone. Further *structures of legitimation* define the appropriate dress code in particular settings, but the transgression of these may invoke sanctions, although often these are covert.

Technology, from Giddens's perspective, cannot be an agent. Thus, a phone is not an agent, but it does have structural properties and 'does nothing, except as implicated in the actions of human beings' (Giddens and Pierson, 1998). However, there have been many suggestions (e.g., Turkle, 2005, 2011) that computers are not merely objects that make our lives more efficient but are subjects that are intimately and, ultimately, linked to our social and emotional lives. Turkle argues then that the result is that computers change not only what we do, but also how we think about the world and ourselves. What is particularly important here is the structure-agency debate that introduces questions about the nature of social behaviour: whether it is ultimately predictable in terms of the creative volition of the individual, or is largely a product of socialization, interaction and greater social structures.

There has been relatively little consideration of agency in relation to virtual humans, whether in 3D worlds or acting as pedagogical agents. Yet agency in 3D worlds is implemented in very novel ways, such as particular activities or functions that can be scripted to make avatars respond in particular, and even autonomous, ways. Therefore, it is a challenge to extend the simple author/avatar relation to a broader consideration of agency since it is reconstituted by the multiple relations between author/avatar/agent/world. There is also a link between the autonomy of the virtual human, as described by Parasuraman and Sheridan's model discussed in Chapter 5, and the agency of the virtual human. What level of autonomy is required for the virtual human to have true agency?

Giddens (1984) suggested human agency and social structure are in a relationship with each other, and it is the repetition of the acts of individual agents which reproduces the structure, but it is not clear how structure and agency relate to one another when working with virtual humans. Much of the work in this area in relation to virtual humans has concentrated on Phenomenological agency, Neurocognitive agency and Attributed agency.

Phenomenological Agency

Phenomenology seeks to explore the human experience at a fundamental level, examining the essence of lived experiences as it is for several individuals. In particular, phenomenologists want to know about the very structures of consciousness. In the context of agency, a phenomenological distinction has been made between the 'feeling of agency', which refers to the feeling of being in control of an action, and the judgement of agency. The judgement of agency describes the ability to make judgments about one's sense of agency and is seen as a higher conceptual level. Whilst this distinction is generally accepted, Saito et al. (2015) have argued that the distinction between explicit judgement of agency and implicit feeling of agency is theoretical. The study examined explicit and implicit agency measures in the same population and found that they did not correlate, suggesting dissociation of the explicit judgement of agency and the implicit feeling of agency. For a virtual human, both the feeling of agency and the judgement of agency would require a relatively high degree of self-awareness in order to be manifest – although an observing physical human may well ascribe or judge the virtual human as having agency without it, or on its behalf.

Neurocognitive Agency

Tsakiris et al. (2007) explored how multisensory signals interacted with body representations to generate the sense of body-ownership, and how the sense of agency affects and moderates the sense of body-ownership. However, Limerick et al. (2014) argue that there are two theoretical views regarding the neurocognitive processes underlying the sense of agency:

1. that the sense of agency arises principally from internal processes serving motor control, and
2. that external contextual cues are important.

Again, a virtual human would need a well-developed sense of self in order to feel such neurocognitive agency, although the concept of a proprioceptive sense has been discussed in Chapter 4. A virtual human developed through exploratory learning approaches could be expected to have a higher sense of neurocognitive agency than one which has been created though a more explicit programming approach. What is perhaps more useful than a clear delineation of phenomenological agency and neurocognitive agency is attributed agency.

Attributed Agency

Attributed agency is both the sense of agency and the way in which agency is attributed to a human or virtual human. In the case of virtual humans, what seems to be evident is that the context affects the sense of agency. For example, Obhi and Hall (2011) found that humans consider face-to-face shared action with other humans different from human-computer shared actions. The findings indicated that attributed agency tends to be overruled when the participant is aware the computer is a co-actor. Studies such as this seems to imply that humans are more likely to attribute agency to other humans but not to virtual humans – even if they are doing the same task, although it could be suggested that this depends upon the type and efficacy of the virtual human. For example, Reeves and Nass (1996) argued that humans do relate to virtual humans in similar ways to humans, but that voice and affect are central to agency attribution. Certainly, work by Guadagno et al. (2007) suggested that agency was affected by the gender match between the human and the virtual human. Furthermore, recent work in this area would support this (Savin-Baden et al., 2013). The study also indicated the importance of the need for behavioural (as opposed to photorealistic) authenticity of the virtual human, in terms of it being effective in influencing humans.

Attributed agency also relates to the ways in which we use technology and the impact it is perceived to have on our agency, namely the affordances of technology.

AGENCY AND AFFORDANCES

Many writers and researchers have suggested that technology has 'affordances', the idea that technology is not neutral but value-laden. In terms of virtual humans, the implication is that they can alter the ways in which we relate to them, and can evoke different views of the world. Postman (1993, 13) has argued:

> Embedded in every tool is an ideological bias, a predisposition to construct the world as one thing rather than another, to value one thing over another, to amplify one sense or skill or attitude more loudly than another ... New technologies alter the structure of our interests: the things we think about. They alter the character of our symbols: the things we think with. And they alter the nature of community: the arena in which thoughts develop.

The concept of affordances has become increasingly used in research and technology since the late 1980s. The term originated from Gibson (1979, 1), who developed the ecological approach to visual perception in which he argued that:

> When no constraints are put in the visuals system, we look around, walk up to something interesting and move around it so as to see it from all sides, and go from one vista to another. That is natural vision …

Thus, is it possible to see how this term has been (mis-)appropriated when it is realized that he argued: 'The *affordances* of the environment are what it offers, the animal, what it *provides* or *furnishes*, for good or ill' (Gibson 1979, 115; original italics). The use then of 'affordances' seems, at one level, to have provoked an overemphasis on what particular technologies prompt or allow us to do, bringing with it a sense of covert control. The underlying assumption here then would be that virtual humans and virtual personas, in particular, could develop covert control of humans, and links with theories of post-humanism. Post-humanism, which seeks to break down binary distinctions between 'human', 'machine' and 'text' (Hayles, 1999, 2012) and between 'nature' and 'culture', also rejects dualisms that are used to define 'being', such as subject/object. Thus, it is theory that is used to question the foundational role of 'humanity' and prompts consideration of what it means to be a human subject, and the extent to which the idea of the human subject is still useful. Whilst it is important to be aware of advertising on social media and in internet games, it is also vital to remember that we have *choice*. We can choose to turn off advertising, we can use internet ad blockers, we can refuse to be coerced. Further, Giddens's view of human agency is that the will rather than the intellect is the ultimate principle of reality. He argues that, apart from instances where someone has drugged or interfered with by others, there is always 'the possibility of doing otherwise' (Giddens, 1979, 258).

PROXEMICS

Proxemics is the study of spatial distances between individuals in different cultures and situations and was first defined by Hall (1996), who argued that humans have an innate distancing mechanism which helps regulate social situations. His model of proxemics described the interpersonal distances of humans (the relative distances between people) in four distinct zones: (1) intimate space, (2) personal space, (3) social space and (4) public space, as depicted in Figure 10.1.

Yee et al. (2007) developed a script to measure the proximal distances and gaze orientations of the 16 avatars closest to the researcher in a virtual

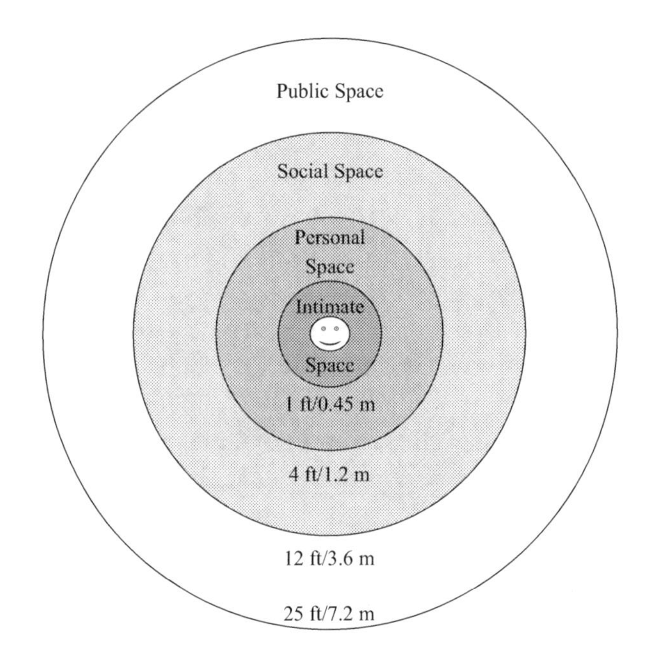

FIGURE 10.1 Hall's interpersonal distances. (After Hall, S., Introduction: Who needs 'identity'? in Hall, S., and du Gay, P. (Eds.), *Questions of Cultural Identity*, Sage Publications, London, UK, 1996.)

200 meter radius, and to track whether or not the observed avatars were talking to each other or not. Their findings suggest that by measuring interpersonal distance, social interactions in virtual worlds like Second Life are governed by the same social norms as those in the physical world. However, an example of the use of proxemics in virtual worlds and mixed reality performance suggests that it perhaps more complex than Yee et al. suggest. Chafer and Childs (2008) deconstructed two scenes from *Hamlet* that were performed live in a recreation of the Globe Theatre in Second Life. They found that issues, such as proxemics, representation and understanding rules, gestures and conventions were all issues they had not initially realized would have such an impact on virtual performance. Savin-Baden (2013) found that ownership, spatial violation and replication were concerns raised by participants in relation to spatial practice. However, in terms of proxemics, participants suggested that an understanding of social cues, spatial negotiation and spatial consideration were important considerations for effective teaching in Second Life.

A more recent study, Lee et al. (2018), investigated proxemics in the presence of a real or virtual human in augmented reality (AR). They sought to raise the mismatch between a small augmented visual field and a large unaugmented periphery by restricting the field of view to the central region. The study results show objective benefits for this approach in producing behaviours that more closely match those that occur when seeing a real human, but also some drawbacks in overall acceptance of the restricted field of view. What was particularly useful in terms of virtual human research were the differences found in the effects that real and virtual humans have on locomotion behaviour in AR with respect to clearance distances, walking speed and head motions.

Identity and agency in virtual spaces may be different in that particular spaces, objects and activities cause avatars to respond in particular ways. This would seem to suggest an opportunity to challenge us to extend the simple author/virtual human relationship to a broader consideration of agency, as it is reconstituted by the multiple relationships between author/virtual human relationship/digital context.

AGENCY AND IDENTITY

Recent studies in higher education also shed light on the issues of agency and identity. Studies in the U.S. education system explored the social interactions of virtual humans and students, considering the humanistic qualities of agent-student interaction and, in particular, the realism of the virtual humans and agent appearance. The focus has thus been on the development of agents that are humanlike and can complete tasks in an efficient manner, in effect, issues of design. Successful learning and engagement, according to Kim and Baylor (2015) seems to be related to the extent to which there is a perceived relationship or association between the Pedagogical Agent and the user:

Appearance: Learners tend to be more influenced by a Pedagogical Agent of the same gender as them and ethnicity, similar to human–human interactions where humans are more persuaded by members of their in-group. Yet, both the learning/motivational context, age of learners and topic play a significant role. One of the issues about appearance is also that of morphology, particularly in relation to gender, race and ethnicity. Until fairly recently, avatars in virtual worlds have been young, thin and Caucasian. Robertson (2010) has raised questions about the over-feminization of the platforms in relation to the fembots in Japan and; it is also notable that there are very few black virtual humans.

Attitude: Using the Pedagogical Agent as a motivator that demonstrates positive attitudes towards the task and the desired levels of performance seems to be helpful for learners to cope in situations where they feel themselves to be novices.

Interaction Model: Using the Pedagogical Agent as a guide or friend seems to be more effective than using one that is perceived to be a high-level expert. Where a Pedagogical Agent acts as a guide, users indicate significantly enhanced motivational outcomes as compared to the mentor who is an expert.

Perceived Competence: Pedagogical Agents with similar competency to learners were more influential than highly competent agents in enhancing student self-efficacy beliefs.

CONCLUSION

It is evident that the research into identity and agency in virtual human research remains complex and inconclusive. Whilst it is possible to define different types of identities, both real-life and in terms of virtual humans, this does not always assist in design and use. Agency remains a complex and contested concept, but understanding the impact of different types of agency, such as Phenomenological agency, Neurocognitive agency and Attributed agency is important for the design of virtual humans. However, what does seem to be evident, though, across studies, is that appearance does matter more than early studies have implied, and that behavioural authenticity is also important for coherence. Appearance, emotion, authenticity and ubiquity are clear components that are important for design across which ever form of virtual human is being developed. This has implications for not only the social and commercial use of virtual humans, but the value of educational impact, which will be explored next in Chapter 11.

REFERENCES

Bernstein, B. (1996). *Pedagogy Symbolic Control and Identity*. London, UK: Taylor & Francis Group.

Chafer, J., & Childs, M. (2008). The impact of the characteristics of a virtual performance: Concepts, constraints and complications. In *Proceedings of ReLIVE08 Conference*, Milton Keynes, UK, November 20–21. Available online www.open.ac.uk/relive08/.

Floridi, L. (2011). The construction of personal identities online. *Minds and Machines, 21*, 477–479.

Furlong, J., & Davies, C. (2012). Young people, new technologies and learning at home: Taking context seriously. *Oxford Review of Education: The Educational and Social Impact of New Technologies on Young People in Britain, 38*(1), 45–62.

Gibson, J. (1979). *The Ecological Approach to Visual Perception*. Boston, MA: Houghton, Mifflin Company.

Giddens, A. (1979). *Central Problems in Social Theory: Action, Structure and Contradiction in Social Analysis*. London, UK: Macmillan.

Giddens, A. (1984). *The Constitution of Society*. Cambridge, UK: Polity Press.

Giddens, A., & Pierson, C. (1998). *Conversations with Anthony Giddens*. Cambridge, UK: Polity Press.

Guadagno, R. E., Blascovich, J., Bailenson, J. N., & McCall, C. (2007). Virtual humans and persuasion: The effects of agency and behavioral realism. *Media Psychology, 10*, 1–22.

Hall, S. (1996). Introduction: Who needs 'identity'? In S. Hall & P. du Gay (Eds.), *Questions of Cultural Identity*. London, UK: Sage Publications.

Haraway, D. (1985). A Cyborg Manifesto: Science, technology, and socialist-feminism in the late twentieth century. In *Simians, Cyborgs and Women: The Reinvention of Nature* (pp. 149–181). New York: Routledge.

Hayles, K. (1999). *How We Became Posthuman: Virtual Bodies in Cybernetics, Literature and Informatics*. Chicago, IL: University of Chicago Press.

Hayles, K. (2012). *How We Think: Digital Media and Contemporary Technogenesis*. Chicago, IL: University of Chicago Press.

Ito, M., Baumer, S., Bittanti, M., Boyd, D., Cody, R., Herr-Stephenson, B. et al. (2010). *Hanging Out, Messing Around, and Geeking Out*. Cambridge, MA: MIT Press.

Kim, Y., & Baylor, A. L. (2015). Research-based design of pedagogical agent roles: A review, progress, and recommendations. *International Journal of Artificial Intelligence in Education, 26*(1), 160–169.

Lee, M., Bruder, G., Höllerer, T., & Welch, G. (2018). Effects of unaugmented periphery and vibrotactile feedback on proxemics with virtual humans. *AR IEEE Transactions on Visualization and Computer Graphics, 24*(4), 1525–1534.

Limerick, H., Coyle, D., & Moore, J. W. (2014). The experience of agency in human-computer interactions: A review. *Frontiers in Human Neuroscience, 8*, 643.

Nakamura, L. (2010). Race and identity in digital media. In J. Curran (Ed.), *Mass Media and Society*. London, UK: Bloomsbury Academic.

Obhi, S. S., & Hall, P. (2011). Sense of agency in joint action: Influence of human and computer co-actors. *Experimental Brain Research*, 211, 663–670.

Postman, N. (1993). *Technopoly: The Surrender of Culture to Technology*. New York: Vintage Books.

Reeves, B., & Nass, C. (1996). *The Media Equation: How People Treat Computers, Television and New Media Like Real People and Places*. New York: Cambridge University Press.

Robertson, J. (2010). Gendering humanoid robots: Robo-sexism in Japan. *Body & Society, 16*(2), 1–36.

Ryberg, T., & Larsen, M. C. (2008). Networked identities: Understanding relationships between strong and weak ties in networked environments. *Journal of Computer Assisted Learning, 24*(2), 103–115.

Saito, N., Takahata, K., Murai, T., & Takahashi, H. (2015). Discrepancy between explicit judgement of agency and implicit feeling of agency: Implications for sense of agency and its disorders. *Conscious and Cognition, 37,* 1–7.

Savin-Baden, M. (2013). Spaces in between us: A qualitative study into the impact of spatial practice when learning in Second Life. *London Review of Education, 11*(1), 59–75.

Savin-Baden, M. (2015). *Rethinking Learning in an Age of Digital Fluency Is being Digitally Tethered a New Learning Nexus?* London, UK: Routledge.

Savin-Baden, M., Tombs, G., Burden, D., & Wood, C. (2013). 'It's almost like talking to a person': Student disclosure to pedagogical agents in sensitive settings. *International Journal of Mobile and Blended Learning, 5*(2), 78–93.

Steils, N. (2013). Exploring learner identity in virtual worlds in higher education: Narratives of pursuit, embodiment and resistance, PhD thesis, Coventry University.

Tsakiris, M., Schütz-Bosbach, S., & Gallagher, S. (2007). On agency and body-ownership: Phenomenological and neurocognitive reflections. *Consciousness & Cognition, 16,* 645–660.

Turkle, S. (1996). *Life on the Screen: Identity in the Age of the Internet.* London, UK: Weidenfeld & Nicolson.

Turkle, S. (1999). Looking toward cyberspace: Beyond grounded sociology. *Contemporary Sociology, 286,* 643–648.

Turkle, S. (2005). *The Second Self: Computers and the Human Spirit* (2nd ed.). Cambridge, MA: MIT Press.

Turkle, S. (2011). *Alone Together.* New York: Basic Books.

Yee, N., Bailenson, J. N., Urbanek, M., Chang, F., & Merget, D. (2007). The unbearable likeness of being digital: The persistence of nonverbal social norms in online virtual environments. *CyberPsychology & Behavior, 10*(1), 115–121.

Žižek, S. (1999). The Matrix, or two sides of perversion. *Philosophy Today,* 43. Available online http://www.nettime.org/Lists-Archives/nettime-l-9912/msg00019.html.

CHAPTER 11

Virtual Humans for Education

INTRODUCTION

This chapter will examine the current and possible future impact that virtual humans could have on education and draw on recent studies undertaken in this area by the author that have implemented virtual humans in different ways in educational settings. The chapter will examine the potential impact of virtual humans on education in terms of new developments and uses of virtual humans for guiding and supporting learning, as well as exploring their impact on our everyday lives, and even on our relationships.

LEARNING

Whilst theories of learning have never been static, the distinction between and across the approaches – behavioural, cognitive, developmental and critical pedagogy – continues to be eroded. There is increasing focus in the twenty-first century on what and how students learn and on ways of creating learning environments to ensure that they learn effectively – although much of this remains contested ground. New models and theories of learning have emerged over the last decade that inform the development of learning in virtual worlds, such as Second Life, and may have an impact on virtual humans.

The work of Trigwell et al. (1999) on teachers' conceptions of learning offers useful insights into the impact such conceptions have on student learning. Yet, the work of Meyer and Land (2006), Haggis (2004) and Meyer and Eley (2006) have been critical of studies into conceptions of teaching and approaches to learning. At the same time, learning related to human-computer interactions has been the focus of much debate (e.g., Turkle, 1996, 2005; Žižek, 2005). Those such as Pirolli (2007) have argued that humans have limited ability to store information, seeming to imply that learning is about gaining knowledge or finding the right information. Yet approaches such as activity-led learning, collaborative learning and high-level constellations of problem-based learning online (Savin-Baden, 2007) remain unrealized by many staff who focus on teaching design rather than learning design. Teaching design focuses on what knowledge and content staff want to teach students. Learning design focuses on what it is students need to learn (which includes a range of capabilities and knowledge) to become, for example, a good engineer or midwife.

One straightforward way to understanding the differences in these learning approaches is to consider their key concepts and how these can have an impact on the design and use of virtual humans, as shown in Table 11.1.

TABLE 11.1 Learning Approaches and Virtual Humans

The Approach	Key Concepts	Learning Challenges	Appropriate Virtual Human
1. The Behavioural Approach	Specific goals and clear objectives are needed for learning. The learning experience should be task orientated.	The focus is on incentives – which do not motivate everyone. The assumption is that having passed the test you can do the job.	Pedagogical agent that focuses on content coverage and testing knowledge outcomes.
2. The Humanistic Approach	The learning needs to be controlled by us as the learners, not by the tutor. Emphasis must be on our freedom to choose the approach.	Too much freedom can be disabling. People are not always sure what they want or need to learn.	Virtual mentor and life coach where the learning is personalised.

(*Continued*)

TABLE 11.1 (*Continued*) Learning Approaches and Virtual Humans

The Approach	Key Concepts	Learning Challenges	Appropriate Virtual Human
3. The Cognitive Approach	Everyone has their own cognitive structure which must be accommodated. People can only learn new information in relation to what we already know.	Overemphasis on learning approaches at the expense of content. Tendency to categorise people into 'types' of learners.	Knowledge expert designed to fit with peoples' approaches to learning and motivations.
4. The Developmental Approach	Learning needs to be part of our progressive development. Knowledge acquisition needs to be relevant in time and context.	Overemphasis on experiential approaches to learning at the expense of effectiveness. Tendency to spend time reflecting on mistakes rather than looking forward.	Virtual mentor and life coach where the learning is designed with the people and adapted according to personal needs.
5. The Critical Awareness Approach	Everyone has values – including tutors. Learning is not value free. Learning always takes place in a social and cultural context.	Difficult to manage power relations between teacher and learner to ensure 'real' equality. Can be seen as overly politicized.	Virtual mentor that focuses on exploring assumptions, hidden agendas and political contexts.

NETWORKED LEARNING AND TECHNOLOGY-ENHANCED LEARNING: SOME DEBATES

To date, there are a wide range of terms that are used about learning in digital settings, few of which include discussions about the use of virtual humans. Invariably, these differences are not defined, yet they do have an impact on learning. This section defines the main terms used and illustrates how they affect the use of virtual humans in the classroom.

Networked Learning

An early definition of networked learning is 'learning in which information and communication technology (ICT) is used to promote connections: between one learner and other learners; between learners and tutors; between a learning community and its learning resources' (Goodyear et al., 2004, 1). This definition of networked learning relates to the networked connections between resources, learners, teachers and learning communities. Over the last 10 years, it has broadened and developed as an area of research and pedagogy and includes exploration into formal and informal learning settings. Networked learning takes a critical stance towards practice, theory, pedagogy, learning design and intuitional agendas. The most recent focus has been on examining relational and interactional aspects of learning and development and dialogic learning in the contexts of social networks.

Technology-Enhanced Learning

Technology-enhanced learning tends to take a quite instrumental approach so that technology is used as a means of supporting institutional and learning needs, thereby separating the social domain. Technology-enhanced learning (TEL) is seen as a series of practices that have swept across higher education with relatively little critique, along with other terms such as engagement and quality. Certainly, the number of projects funded by the European Union over the last 10 years would seem to suggest this is the case. Bayne (2014) has argued that technology-enhanced learning is more about technology than learning. She argues that TEL is really about neither enhancement nor learning, since TEL merely focuses on the instrumental. The result of adopting this approach then is a lack of critique of the impact of technology on pedagogy and the possible impact it may have on discipline-based pedagogy.

Online Learning

Online learning is generally seen as learning that takes place using digital media, but, generally, it is learning that takes place at a distance. Thus, students study a course or program from home or work, often through some kind of learning portal management system at a university, but students don't need to attend university sessions face to face. Students can study for recognised qualifications without needing to attend classes on campus, and it is aimed at and designed for those who wish to study for a qualification alongside work or other commitments.

Blended Learning

Blended learning is online education that is combined with face-to-face education, and it is often seen as a means of learning more efficiently, for instance by removing barriers of time and distance. It may also enable the use of novel instructional methods and make education more student-centred. In the past, distance-based learning activities have often been associated with traditional delivery-based methods, individual learning and limited contact.

Different dimensions of 'blending' have been identified, such as blending instructional modalities, blending delivery media or blending instructional methods. The term blended learning is sometimes also used to refer to the use of technology in face-to-face education, but it is suggested here that blended learning is 'a combination of traditional face-to-face and online instruction' (Graham, 2013, 334).

Digital Education

Digital education is active learning in digital spaces which encourages digital fluency and enables learning that students would be unable to experience in other ways. Digital education, also known as eLearning, needs to be seen as embedded across geographies as well as physical and virtual realms so that home/school/work/leisure/play become spaces of learning criticality and creatively.

The views of teachers and students about the value of digital education still differs markedly in many schools and universities. Thus, embedding digital education effectively will improve student learning and engagement since students struggle with traditional, or lecture-based, forms of learning.

Table 11.2 summarises these different forms of eLearning and highlights their challenges and implications for virtual humans.

TABLE 11.2 Forms of eLearning

	Definitions	**Challenges**	**Implications for Virtual Humans' Use**
Networked learning	Learning that uses technology and media to connect resources, learners, teachers and learning communities.	It is an approach that challenges the status quo and therefore can be difficult to implement or gain acceptance at an institutional level.	Difficult to implement in sensitive settings.
Technology-enhanced learning	The use of technology to support institutional learning.	Implementation can often be narrow and highly managed with a focus on technology over pedagogy.	Only narrow types of virtual humans can be used, with guided learning, such as pedagogical agents.

(Continued)

TABLE 11.2 (*Continued*) Forms of eLearning

	Definitions	Challenges	Implications for Virtual Humans' Use
Online learning	Learning that occurs in digital spaces, often in a designated learning environment such as a learning management system.	Technology failure can result in students feeling unsupported or unsure about how to use the technology.	Virtual humans need to be well designed with built-in support and an online 24/7 help desk.
Blended learning	Online education that is combined with face-to-face education.	Linking face-to-face and online learning can result in there being a devaluing of learning online over face-to-face sessions.	Virtual humans need to be well designed to match both face-to-face and online contexts.
Digital education	Active learning in digital spaces, which encourages digital fluency and enables learning that students would be unable to experience in other ways.	Embedding it beyond a Virtual Learning Environment (VLE) engaging staff, ensuring active and collaborative forms of learning.	Virtual humans become part of the education system that prompts and promotes digital fluency and improves student learning and engagement.

VIRTUAL HUMANS AS INTERRUPTION AND CHANGE TO CURRENT PEDAGOGIC PRACTICES

One of the challenges of introducing virtual humans in the guise of virtual tutors, coaches or pedagogic agents into education is that it often results in unwanted change. One of the worries across education is that virtual humans may make staff redundant and control future employment. There is also the assumption that if there are more teachers in the classrooms that students will learn more effectively. Although, in general, universities do not force the use of virtual tutors on staff or students, the use of them across an institution can result in questions being asked both about the innovation being introduced and also as to why change is required. Perhaps one of the most striking things about the introduction of virtual tutors is that it does force staff to reconsider:

- The use of current teaching practices and their relevance to a net generation.
- The extent to which their current practices engage students effectively.

- Whether virtual tutors can improve or add value to learning.
- Whether the use of virtual tutors is just another means of providing infotainment for students, which prevents a critical stance towards knowledge.

Yet, staff stances towards these types of virtual humans remains quite polarised between seeing them as a threat or seeing them as a useful means of developing learning and teaching. Bayne and Jandric (2017, 111) argue:

> What kind of combination of human and artificial intelligence will we be able to draw on in the future to provide teaching of the very best quality? What do we actually want from artificial intelligence? We should not allow artificial intelligence in education to be driven entirely by corporations or economists or computing and data scientists – we should be thinking about how we take control as teachers. So, the important questions to be asked are: How could we do our jobs better with artificial intelligences? What might that look like, and how might our students benefit?

Staff and students need to consider how it might be useful to work in partnership with virtual humans and what new kinds of learning might be brought in to schools and universities. The main challenge of using virtual humans in higher education would seem to be in helping staff and students to ways in which they might use virtual humans for learning.

At the other end of the scale, toys, tracking and learning today are perhaps more intertwined than many people would envisage and, therefore, it is important to understand them together. One such example is the use of chatbot-enabled Barbie Dolls (Vlahos, 2015). Tracking, whether through apps, sports watches or online shopping, is now commonplace, and whilst many people are aware of this tracking, it is not (yet) perceived to be malicious or sinister. As consumers, both now and in the future, children are being influenced in their buying choices and guided to other retail spaces through related online marketing. These technologies influence early learning in the home and at preschool, but as Plowman (2014) notes, children, and particularly preschool children, are largely invisible in studies about family life. The result is that there is still relatively little knowledge and understanding about how young children learn and are influenced by the wide range of connected, and often unsupervised, technologies in the home.

USING VIRTUAL HUMANS IN EDUCATIONAL SETTINGS

In terms of young adults, and students in particular, evidence suggests that pedagogic agents, virtual tutors and virtual mentors have a role to play in the support of student and staff education, learning, access to information and personal support. For example, one recent development was the creation of an automated teacher at the University of Edinburgh (Bayne, 2015) in the context of a Massive Open Online Course (MOOC). According to Bayne (2015, 461), this Twitterbot:

> 'Coded in' something of the teacher function to the MOOC, using it as a way of researching some creative and critical futures for a MOOC pedagogy in which the 'teacher function' might become less a question of living teacher presence and more an assemblage of code, algorithm and teacher–student agency.

There are a range of uses for virtual humans within an education setting, such as those being used as teachable agents and teaching assistants and those that are used to motivate students.

Virtual Humans as Teachable Agents

A Teachable Agent (TA) is a form of Pedagogical Agent that builds upon the pedagogy of *learning by teaching* (Bargh and Schul, 1980; Chase et al., 2009). These kinds of agents allow the learner to teach the TA about the content and relationships regarding a particular topic. Furthermore, TA can then support the learning of the human user by posing questions based on the taught content. Although it is clear to the learner that the agent is not a real person, the agent can be viewed as both a peer or student and also as a teacher/mentor. One key area of importance here links to the coding of knowledge described earlier in this review. An individual's attempts at teaching the TA requires that the individual records their personal understanding and relationships of the topic into the electronic medium that then guides future responses by the agent.

Virtual Humans as Teaching Assistants

There have been a few recent studies that have developed virtual humans in order to help students to develop specific capabilities, such as assessment and negotiation skills, as well as the extent to which virtual humans affect and influence behaviour and learning.

Gratch et al. (2015) created a virtual human to enable students to learn negotiation skills. The authors suggest that negotiation demands the development of three key skills in virtual humans: intelligence, language and embodiment. In practice, the researchers created a conflict resolution agent in a game-type environment where students were presented with a wide range of dispute resolution concepts that mirror the kinds of negations games used in business schools. The authors argue that using virtual humans allows students to tailor their experience to match their current skills and receive targeted feedback. The findings of the evaluation provided some useful guidance for future development and suggest that the virtual human evokes similar behaviours that occur in face-to-face negotiations, but no statistically significant differences or benefits were indicated in the findings.

White et al. (2015) used a virtual human to simulate critical incidents with health care professionals. In practice, the study examined how effectively information was transferred from nurses during a medical situation between nurses and a virtual attending physician.

Checklists were used to evaluate information transfer and the findings indicated that nurses were not consistent in the ways information was shared and that many missed a fatal dosage error. What was particularly useful in this study was not only the effective use of a virtual human but also the way in which it highlighted practice errors that could be improved by using virtual humans as critical team members.

Issues such as human error, accuracy and truthfulness are all areas that appear to be ones in which virtual humans can be used. This is exemplified further by a series of studies by Savin-Baden et al. (2013) who undertook a study to evaluate the potential influence of a pedagogical agent in affecting a person's reactions and responses with regards to truthfulness, disclosure and personal engagement, and to use these findings to consider its application in and beyond educational contexts. It was found that, whilst technical realism is important to willingness to disclose information, what one participant experiences as split-attention effect, can be experienced by another as a conversational partner's lack of engagement. The implications of this study are that truthfulness, personalisation and emotional engagement are all vital components in using pedagogical agents to enhance online learning. A later study by Savin-Baden et al. (2015), found that emotional interactions with pedagogical agents were intrinsic to a user's sense of trust, and that truthfulness, personalisation and emotional engagement are vital when using pedagogical agents to enhance online learning. This was confirmed by Savin-Baden et al. (2016) with a study that was undertaken to examine human interaction with a pedagogical agent and the passive and active detection of such agents within a synchronous, online environment. The passive detection test was where participants were not told of the presence of a pedagogical agent within the online environment. The active detection test was where participants were primed about the potential presence of a pedagogical agent. The purpose of the study was to examine how people passively detected pedagogical agents that were presenting themselves as humans in an online environment. In order to locate the pedagogical agent in a realistic higher education online environment, problem-based learning online was used. Problem-based learning online provides a focus for discussions and participation, without creating too much artificiality. The findings indicated that the ways in which students positioned the agent tended to influence the interaction between them. One of the key findings was that since the agent was focused mainly on the pedagogical task, this may have hampered interaction with the students, however, some of its non-task dialogue did improve students' perceptions of the autonomous agents' ability to interact with them. It is suggested that future studies explore the differences between the relationships and interactions of learner and pedagogical agent within authentic situations, in order to understand if students' interactions are different between real and virtual mentors in an online setting.

Goel and Joyner (2016) undertook a study that not only explored the use of virtual humans to motivate learning, but also redesigned a course so that it would work effectively with a 'nano-tutor'. One of the issues in the education design of courses is that current content and knowledge is merely repurposed rather than being designed especially for use with virtual humans. Goel and Joyner (2016) offered an online course entitled 'Knowledge-Based Artificial Intelligence: Cognitive Systems'. They began by characterising the virtual human as human-level, human-centred, and human-like AI. They also elected not to use material previously used by designing a new course with new context and activities. The authors adopted active learning approaches (such as project-based learning and interactive educational technology) and used AI agents as 'nano-tutors' in the interactive exercises of the online course. The research was undertaken using a comparative design using human tutors in parallel with the online course. The findings indicate that the students in the online course performed at least as well as the students in the face-to-face class in the assessments, and students were generally positive about engaging with the virtual human. This was a useful study, but the findings did not provide details about issues, such as emotional engagement, or the impact of the virtual human on learning.

The studies presented above illustrate the ways in which virtual humans can be used for learning in different disciplines and profession spheres. There is also a growing body of literature that is exploring the use of virtual humans to motivate learners in general.

Virtual Humans for Motivating Learning

There are few studies that have explored the use of virtual humans for motivating learning, and most of them have been undertaken in secondary schools. For example, a study by van der Meij et al. (2015) designed and tested a motivational animated pedagogical agent. The authors placed the agent within a kinematics program (the branch of mechanics concerned with the motion of objects) that used inquiry-based learning. The agent was designed to appeal to female students as they are perceived to be underrepresented in science classrooms and deemed to require special attention, and, therefore, the agent was female, 'young, attractive, and "cool"' (van der Meij et al. (2015, 308)) – which does introduce issues, again, about morphology and stereotypes in virtual humans. The study sought to understand if students' motivation and knowledge changed over time and whether gender affected such changes. The findings indicated that with self-efficacy, beliefs increased significantly for both boys and girls halfway during training; girls' self-efficacy beliefs significantly increased in both experimental conditions and decreased in the control condition. However, the knowledge gains and motivation were the same for boys and girls. The authors conclude that in designing an agent that can influence student motivation, a strong focus should be on the design of the agent.

A study by Dincer and Doganay (2015) also explored the impact of a pedagogical agent on secondary school students' academic success and motivation. The study compared four different groups. The first group received education though a pre-decided pedagogical agent, the second group could choose the pedagogical agents, the third group received the education without a pedagogical agent and the final group received the same education through a traditional teaching. The findings of the study support earlier work by Savin-Baden et al. (2013) which suggest that students prefer to choose their own agent, including clothes and voice. However, Dincer and Doganay (2015) also found that allowing students choice and personalisation of the agent led to an increase in student motivation.

LEARNING, IMMERSION AND VIRTUAL HUMANS

Although it has been recognised in both schooling and higher education that a thorough engagement in tasks results in effective learning (Dewey, 1938; Bruner, 1991; Gee, 2004), it is only relatively recently that the notion of immersion has come to the fore. Games such as the Quest Atlantis Project (Barab et al., 2007), a 3D game for children, and the River City MUVE (Galas and Ketelhut, 2006), essentially seem to have embraced immersion as a central component of learning. Whilst immersion has been central to discussion and explorations of virtual reality, immersion has only become of more interest in terms of virtual humans since the development and popularity of virtual worlds. One of the areas of interest raised by staff in a study (Savin-Baden, 2010) was that of immersion and the impact of immersion on learning. Nonetheless immersion, even on a small scale, can be disarming, and staff have raised the issue that anxieties emerged when difficulties occurred with their avatars, such as feeling in danger, out of control or falling in water. Certainly, a strong feature of learning in virtual worlds is the way in which the user's attention is captivated and results in a sense of immersion or presence (Steuer, 1992; Robertson et al., 1997). For example, Dede (1995) describes immersion within learning environments as the subjective impression that a user is participating in a 'world' comprehensive and realistic enough to induce the 'willing suspension of disbelief' (Coleridge, 1817). Immersion is a complex concept related to the physical senses and mental processes of the user, the required tasks within the environment and the types of interaction and technology involved (Pausch et al., 1997). Yet a highly immersive environment will lead to a sense of the user feeling 'in' or 'part of' a virtual environment as they interact with it and become absorbed or deeply involved. This introduces questions about the context in which virtual humans are placed and the impact this has on learning, as illustrated in the studies presented below.

DIRECTIONS FOR FUTURE RESEARCH

In order to understand how virtual humans can be used effectively for teaching and learning, future advances need to focus on the co-development of the pedagogy and technology. Research also needs to examine the relationships between humans and virtual humans within a range of contexts; not just in the role of teachers or lecturers but also in taking up roles or positions as co-learners. Further research and development is needed in the following areas:

- Improving techniques for designing and encoding the database of responses to natural language inputs;
- Increasing use of automated strategies for acquisition and constructing of databases using technologies such as Neural Networks;
- Using virtual humans in a mentoring, guiding or challenging role rather than as purely a source of factual information;
- Examining the roles that virtual humans could take that move away from a 'teacher' to other roles such as a 'peer', 'friend' or 'mentor' who acts as guide, confidant or advisor as appropriate; and
- Exploring factors that influence and enhance engagement with virtual humans, such as the role and believability of the virtual human.

Liew, Tan and Jayothisa (2013) note the issues of agent design remain a complex area. Early studies focused more on the technology than affective issues (de Rosis et al., 2004), and later studies explored agent stereotyping (e.g., Moreno and Flowerday, 2006; Kim and Wei, 2011). Agents should be designed so that they are:

- Similar to the user,
- Aspirational role models,
- Visually attractive (Baylor, 2009; Gulz and Haake, 2010),
- Influential in enhancing a learner's self-efficacy beliefs (i.e., their belief that they are able to accomplish tasks) (Baylor, 2009), and
- Visually present as a character (Baylor, 2009).

Veletsianos and Russell (2013) argue that the social discourse that occurs between agents and learners is generally overlooked in the educational literature and suggest it is vital to examine explicit and concealed meanings in order to gain in-depth understandings of agent-learner interactions and relationships. The authors argue that well designed agents can ask guiding questions, prompt reflection, provide feedback and summarise information. They also suggest that Pedagogical Agents exhibiting human attributes are capable of participating in high-quality conversations, which may result in helpful outcomes, such as alleviating anxiety (Gulz, 2005; Gustafson and Bell, 2005). The study by Veletsianos and Russell (2013) explored the nature and content of interactions when adult learners were offered the opportunity to communicate with Pedagogical Agents in open-ended dialogue. Data were collected over a four-week period with 52 undergraduate students, using a male and a female agent. Findings revealed six themes (388–389) that described the nature and content of agent-learner interactions:

- Conversations positioned agents in multiple instructional and social roles;
- Pedagogical Agents' relationship status and love interests;
- Playful commentary;
- Working toward understanding;
- Learners asked agents personal questions, but were unresponsive to agent
- requests to talk about themselves; and
- Sporadic on-task interactions with limited follow-up.

These results from Veletsianos and Russell (2013, 397) indicated that:

1. Even if Pedagogical Agents are designed and positioned in a particular role (for example, expert/mentor) students may position them in a different or multiple roles.
2. Attempts to prevent or discourage non-task conversations may be misguided, as the role of off-task and non-task interactions in agent-learner conversations is often seen as problematic, yet, in this study, it was apparent that non-task conversations were used to establish rapport and build relationships.
3. Although students treated the agents in ways that seemed to indicate a form of human-human interaction, they also displayed apathy

about sharing information about themselves. The authors suggest this was a case in which users ignored agent questions, which might signify that the relationship between agents and learners may not be perceived by learners to be equivalent to a human-human relationship.

CONCLUSION

The use of virtual humans in education has great potential for improving the access to information and can be used to enhance the learning process through increases in motivation and possibly through perceived novelty. However, the uses of Pedagogical Agents in education are still relatively unexplored, with a prevalent focus still on technical and design features rather than on the underlying pedagogy of their deployment. As virtual human technologies are increasingly integrated into commercial and educational arenas, it seems likely that they will transfer to mobile, as well as blended learning settings. It is suggested, therefore, that such applications require both pedagogical nuance and further research into the ways in which student perceptions of Pedagogical Agents are informed by the context within which they interact. It is evident that there is a wide range of possibilities for using virtual humans in education, from early years to higher and further education, although many of these areas remain underexplored.

REFERENCES

Barab, S., Dodge, T., Tuzun, H., Job-Sluder, K., Jackson, C., Arici, A., Job-Sluder, L., Carteaux, R., Jr., Gilbertson, J., & Heiselt, C. (2007). The Quest Atlantis Project: A socially-responsive play space for learning. In B. E. Shelton and D. Wiley (Eds.), *The Educational Design and Use of Simulation Computer Games*. Rotterdam, the Netherlands: Sense Publishers.

Bargh, J. A., & Schul, Y. (1980). On the cognitive benefits of teaching. *Journal of Educational Psychology, 72*, 593–604.

Baylor, A. (2009). Promoting motivation with virtual agent and avatars: Role of visual presence and appearance. *Philosophical Transactions of the Royal Society of London B: Biological Sciences, 364*, 3559–3565.

Bayne, S. (2014). What's the matter with 'technology enhanced learning'? *Learning, Media and Technology, 40*(1), 5–20.

Bayne, S. (2015). Teacherbot: Interventions in automated teaching. *Teaching in Higher Education, 20*(4), 455–467.

Bayne, S., & Jandric, P. (2017). From anthropocentric humanism to critical posthumanism in digital education. *Knowledge Cultures, 5*(2), 197–216.

Bruner, J. (1991). *Acts of Meaning*. Cambridge MA: Harvard University Press.
Chase, C. C., Chin, D. B., Oppezzo, M. A., & Schwartz, D. L. (2009). Teachable agents and the protégé effect: Increasing the effort towards learning. *Journal of Science Education and Technology, 18*(4), 334–352.
Coleridge, S. T. (1817). *Biographia Literaria*. Princeton, NJ: Princeton University Press (printed 1983).
de Rosis, F., Pelachaud, C., & Poggi, I. (2004). Transcultural believability in embodied agents: A matter of consistent adaptation. In S. Payr & R. Trappl (Eds.), *Agent Culture: Human-Agent Interaction in a Multicultural World* (pp. 75–106). Mahwah, NJ: Laurence Erlbaum Associates.
Dede, C. (1995). The evolution of constructivist learning environments: Immersion in distributed, virtual worlds. *Educational Technology, 35*(5), 46–52.
Dewey, J. (1938). *Experience and Education*. New York: Collier and Kappa Delta Pi.
Dincer, S., & Doganay. A. (2015). The impact of pedagogical agent on learners' motivation and academic success. *Practice and Theory in Systems of Education, 10*(4) 329–348.
Galas, C., & Ketelhut, D. J. (2006). River city, the MUVE. *Leading and Learning with Technology, 33*(7), 31–32.
Gee, J. P. (2004). *What Video Games Have to Teach Us about Learning and Literacy*. Basingstoke, UK: Palgrave Macmillan.
Goel, A., & Joyner, D. (2016). An experiment in teaching cognitive systems online. *Scholarship of Technology-Enhanced Learning, 1*(1), 3–23.
Goodyear, P., Banks, S., Hodgson, V., & McConnell, D. (2004). Research on networked learning: An overview. In P. Goodyear, S. Banks, V. Hodgson, & D. McConnell (Eds.), *Advances in Research on Networked Learning* (pp. 1–10). Boston, MA: Kluwer.
Graham, C. R. (2013). Emerging practice and research in blended learning. In: M. J. Moore (Ed.), *Handbook of Distance Education* (3rd ed.). New York: Routledge.
Gratch, J., DeVault, D., Lucas, G., & Marsella, S. (2015). Negotiation as a challenge problem for virtual humans. *15th International Conference on Intelligent Virtual Agents*, Delft, the Netherlands.
Gulz, A. (2005). Social enrichment by virtual characters - differential benefits. *Journal of Computer Assisted Learning, 21*(6), 405–418. doi:10.1111/j.1365-2729.2005.00147.
Gulz, A., & Haake, M. (2010). Challenging gender stereotypes using virtual pedagogical characters. In S. Goodman, S. Booth, & G. Kirkup (Eds.), *Gender Issues in Learning and Working with Information Technology: Social Constructs and Cultural Contexts* (pp. 113–132). Hershey, PA: IGI Global.
Gustafson, J., & Bell, L. (2000). Speech technology on trial: Experiences from the August system. *Natural Language Engineering, 6*(3&4), 273–286. doi:10.1017/s1351324900002485.
Haggis, T. (2004). Meaning, identity and 'motivation': Expanding what matters in understanding learning in higher education? *Studies in Higher Education, 29*(3), 335–352.
Kim, Y., and Wei, Q. (2011). The impact of learner attributes and learner choice in an agent-based environment. *Computers & Education, 56*(2), 505–514.

Liew, T. W., Tan, S. M., & Jayothisa, C. (2013). The effects of peer-like and expert-like pedagogical agents on learners' agent perceptions, task-related attitudes, and learning achievement. *Educational Technology & Society, 16*(4), 275–286.

Meyer, J. H. F., & Eley, M. G. (2006).The approaches to teaching inventory: A critique of its development and applicability. *British Journal of Education Psychology, 76*, 633–649.

Meyer, J. H. F., & Land, R. (2006). Threshold concepts and troublesome knowledge: Issues of liminality. In J. H. F. Meyer and R. Land (Eds.), *Overcoming Barriers to Student Understanding: Threshold Concepts and Troublesome Knowledge*. Abingdon, UK: RoutledgeFalmer.

Moreno, R., and Flowerday, T. (2006). Students' choice of animated pedagogical agents in science learning: A test of the similarity-attraction hypothesis on gender and ethnicity. *Contemporary Educational Psychology, 31*(2), 186–207.

Pausch, R., Proffitt, D., & Williams, G. (1997). Quantifying immersion in virtual reality. In *Proceedings of the 24th Annual Conference on Computer Graphics and Interactive Techniques*. New York: ACM Press/Addison Wesley.

Pirolli, P. (2007). *Information Foraging Theory: Adaptive Interaction with Information*. Oxford, UK: Oxford University Press.

Plowman, L. (2014). Researching young children's everyday uses of technology in the family home. *Interacting with Computers, 27*(1), 36–46.

Robertson, G., Czerwinski, M., & van Dantzich, M. (1997). Immersion in desktop virtual reality. In *Proceedings of the 10th Annual ACM Symposium on User Interface Software and Technology*. New York: ACM Press.

Savin-Baden, M. (2007). A *Practical Guide to Problem-based Learning Online*. London, UK: Routledge.

Savin-Baden, M. (2010). Changelings and shape shifters? Identity play and pedagogical positioning of staff in immersive virtual worlds. *London Review of Education, 8*(1), 25–38.

Savin-Baden, M., Bhakta, R., & Burden, D. (2016). Cyber Enigmas? Passive detection and pedagogical agents: Can students spot the fake? In S. J. Cranmer, N. Bonderup-Dohn, M. De Laat, T. Ryberg, & J.-A. Sime (Eds.), *Proceedings of the Tenth International Conference on Networked Learning 2016: Looking Back – Moving Forward* (pp. 456–463). Lancaster, UK: Lancaster University.

Savin-Baden, M., Tombs, G., & Bhakta, R. (2015). Beyond robotic wastelands of time: Abandoned pedagogical agents and new pedalled pedagogies. *E-Learning and Digital Media, 12*(3–4), 295–314.

Savin-Baden, M., Tombs, G., Burden, D., & Wood, C. (2013). It's almost like talking to a person: Student disclosure to pedagogical agents in sensitive settings. *International Journal of Mobile and Blended Learning, 5*(2), 78–93.

Steuer, J. (1992). Defining virtual reality: Dimensions determining telepresence. *Journal of Communication, 42*(24), 73–93.

Trigwell, K., Prosser, M., & Waterhouse, F. (1999). Relations between teachers' approaches to teaching and students' approaches to learning. *Higher Education, 37*, 57–70.

Turkle, S. (1996). *Life on the Screen: Identity in the Age of the Internet*. London, UK: Phoenix.

Turkle, S. (2005). *The Second Self. Computers and the Human Spirit.* Cambridge, MA: MIT Press.

van der Meij, H., van der Meij, J., & Harmsen, R. (2015). Animated pedagogical agents effects on enhancing student motivation and learning in a science inquiry learning environment. *Educational Technology Research and Development, 63*(3), 381–403.

Veletsianos, G., & Russell, G. (2013). What do learners and pedagogical agents discuss when given opportunities for open-ended dialogue ? *Journal of Educational Computing Research*, *48*(3), 381–401.

Vlahos, J. (2015) Barbie wants to get to know your child. *New York Times*, September 16. Available online https://www.nytimes.com/2015/09/20/magazine/barbie-wants-to-get-to-know-your-child.html.

White, C., Chuah, J., Robb, A., Lok, B., Lampotang, S., Lizdas, D., Martindale, J., Pi, G., & Wendling, A. (2015). Using a critical incident scenario with virtual humans to assess educational needs of nurses in a postanesthesia care unit. *Journal of Continuing Education in the Health Professions*, *35*(10), 158–165.

Žižek, S. (2005). *Interrogating the Real.* London, UK: Continuum.

CHAPTER 12

Digital Immortality

INTRODUCTION

One of the most interesting, and potentially far-reaching, applications of virtual human technology is within the concept of digital immortality. This chapter examines the research and literature around active digital immortality and explores the emotional, social, financial and business impact of active digital immortality on relations, friends, colleagues and institutions. This chapter presents recent developments in the area of digital immortality, explores how such immortality might be created and raises challenging issues, but also reflects on the ethical concerns introduced in Chapter 9.

BACKGROUND

Digital immortality is the continuation of an active or passive digital presence after death. It occurs when there is no 'digital death' so that the person's digital presence is continued after death (Sofka et al., 2012). Advances in knowledge management, data mining and artificial intelligence are now making a more active presence after death possible, and the dead remain part of our lives as they live on in our digital devices. Despite media interest in digital immortality, research in this area remains relatively small, with little exploration of the impact of this on grief and mourning. This chapter explores the impact on 'adopters' of digital immortality as they prepare to preserve their soul and assets digitally after death, and on 'recipients' encountering digital legacies, such as relatives, lawyers, politicians, and religious leaders.

The concept of digital immortality has emerged over the past decade and digital immortality has already moved beyond digital memorialization towards a desire to preserve oneself after death and is exemplified in the following fictional scenario:

> *Tom played in his rugby club's junior teams and first XV, before coaching the under 11s. Diagnosed with a brain tumour, and telling only his family, he created his own representative 'coaching' avatar, and then used the club's social media to announce both his illness and launch his avatar. Tom's digital legacy of photos, match videos and coaching avatar remains on the club's social media and website.*

Scenarios like this will occur with increasing frequency in the future, so there needs to be an understanding of and an exploration of the different ways that 'creating' digital immortality results in digital legacies, and the impact this may have on both recipients and wider society.

Recent research and development in this area focus on the wider and more global implications of media death. A study by Birnhack and Morse (2018) undertook a national survey in Israel asking how people wanted to deal with their digital remains. They found that awareness of existing online tools for managing access to digital remains is low and the actual use is limited and that spouses are the preferred fiduciary by most users. Furthermore, as there is no policy in place, between a third and a half of the users will leave access by default to the person who retains the personal devices. Cuminskey and Hjorth (2018) explore representing, sharing, and remembering of loss. They suggest that mobile media result in a form of entanglement with the dead that are both private and public.

What is of both interest and concern is who owns the data, what the relationship is between privacy and commemoration and whether since there are so few guidelines there needs to be an etiquette about how death online is managed. Certainly, new work in this area, such as that by Kasket (2019), introduces queries about the unintended consequences of the digital on death, suggesting that the dead continue to 'speak' themselves. Further she asks pertinent questions about how we manage our own data and what might occur when corporations compete with us for control of these data?

It seems that with these studies and developments that there is an increasing interest in the management of online death and that despite much of the current software being rather an embellishing of the true possibilities (as discussed below), there is a growing fascination and demand for digital immortal creation. For example, the use of virtual assistants, such as Siri, that provide voice and conversational interfaces, the growth of machine learning techniques to mine large data sets, and the rise in the level of autonomy being given to computer-controlled systems all represent shifts in technology that are enhancing the creation of digital immortality. The growth of personality capture, mind uploading and levels of simulation, as well as computationally-inspired life after death, may change the future of religion, affect understandings of the afterlife and increase the influence of the dead surviving in society.

RESEARCH AND LITERATURE

There are already companies dedicated to creating digitally immortal personas (Eternime, 2017; Lifenaut Project, 2017); Facebook has now put in place measures to control the post-mortem data on their site (Brubaker and Callison-Burch, 2016). Steinhart (2014) has examined personality capture, mind uploading, and levels of simulation, arguing for a computationally-inspired theory of life after death that will change the future of religion radically. However, there is a need for clarity in the lexicon being used (Bassett, 2015, 2017) and an understanding of how the dead, despite being dead, may still survive in society (Walter, 2017).

The idea of being able to live on beyond your natural death has a long history in human culture and remains popular in novels such as *The Night's Dawn Trilogy* by Peter Hamilton. Prior to our technological age, the agency for this was typically the ghost, and in recent times, we have seen examples from *Randall and Hopkirk* (*Deceased*) to *Ghost* and *Truly, Madly, Deeply*. However, in the digital era, most of the AI within science fiction have tended to be 'evolved' artificial intelligence, such as SkyNet in *Terminator* and Ultron in the *Marvel* films, which have become sentient rather than being created as digital immortal personas of other people. However, Chapter 2 has already discussed digital immortals, or at least the digital copies of real but dead people, in series such as *Caprica* and *Planet B*. What is interesting about these examples, though, is that the digital persona is very much living in the here and now of their progenitor's death, rather than facing up to the implications of potential immortality.

MODELS, CONCEPTS AND PRACTICES

Despite the range of studies into grief and mourning in relation to the digital, research to date largely focuses on the cultural practices and meanings that are played out in and through digital environments, for example:

Models of Grief

There are a number of models of grief, ranging from stage models, such as Kubler-Ross (1969) and Bowlby (1981), to more recent models, defined below, which would seem to have greater resonance with the idea of digital immortality. These include:

Continuing bonds: Throughout most of the twentieth century, it was expected that grieving involved a process of letting go, breaking bonds with the deceased. The continuing bonds model (Klass et al., 1996) argues that, in fact, the relationship with the deceased changes, rather than breaks. In short, the relationship does not end.

Dual process grieving: This model of grieving (Stroebe and Schut, 1999) suggests that there are two categories of stressors affecting bereavement: loss orientation and restoration orientation. Loss orientation refers to the way the bereaved person processes the actual loss and the restoration orientation refers to the secondary consequences of bereavement, such as dealing with the world without the deceased person in it.

Growing around grief: This model suggests that grief does not necessarily disappear over time, but instead of moving on from grief, people 'grow around it' (Tonkin, 2012).

Digital Grief Concepts

These are concepts that have developed in order to make sense of the ways in which digital technology is being harnessed to commemorate and memorialize the dead:

Digital persistence: Kasket (2019) argues that online persistence and the ongoing presence of the data of the dead online will lead to more of a globalized, secularized ancestor veneration culture, and it is important to recognise the ongoing persistence of the dead online on social media, such as LinkedIn, Amazon and YouTube.

One-way immortality and two-way immortality: The former is where the creator has a digital presence in which an individual's in-life profile has been put into memorialized/remembering status in places, such as Facebook. Two-way immortality is where there is the potential for the creator to interact with the living world; this interaction could come in a wide variety of ways, from two-way text or even voice and video conversations by creating a robot using old texts, for example the Griefbot (Bridge, 2016).

The restless dead: Nansen et al. (2015) argue that forms of digital commemoration are resulting in cultural shifts towards a restless posthumous existence. Thus, there is a shift away from the idea of death as being sleep or rest (Hallam and Hockey, 2001), towards the restless dead as they materialize through social media and technical capabilities. Such media include living headstones, digitally augmented coffins and commemorative urns embodying the head of the deceased, and which are seen to interrupt the previous limitations of cemeteries, static headstones and biological death.

Digital Grief Practices

These are particular grief practices that have developed through media and digital media that have become acceptable norms:

Media mourning: Media mourning is defined here as the idea that we are urged to mourn something that is not our grief through social media, such as the 2017 Manchester Arena bombing, or to mourn our personal loss through social media in a highly public way.

Durable biography: Walter (1996) argues that the purpose of grieving is to construct a durable biography that allows survivors to continue to integrate the deceased person into their lives and to find a stable and secure place for them. In practice, this now tends to occur more often through digital memorization and the use of companies such as Eternime.

Virtual veneration: This is the process of memorializing people through avatars in online games and 3D virtual worlds. An example of this are ancestor veneration avatars (Bainbridge, 2013), that are used as a medium for memorializing the dead, and exploiting their wisdom by experiencing a virtual world as they might have done so.

Digital will creation: It is increasingly common for people to create a digital will in which they indicate what is to be done with their digital legacy and assets, and includes passwords and security questions.

This range of models, concepts and practices, illustrates the need to explore cultural meanings and performance of grief within digital environments.

FORMS OF DIGITAL IMMORTALITY

These recent developments would seem to suggest socio-political impacts, such as shifts in understandings about embodiment, death and afterlife, new perceptions of the social life of the dead and new forms of post-mortem veneration. There needs to be an understanding of what seems to be the emergence of different ways of 'creating' digital immortality that results in digital legacies and the impact these may have on recipients and the wider impact they may have on the rest of society. There appear to be different forms of digital immortality in terms of those who create and those who are on the receiving end of digital immortality, as well as the digital legacy left behind:

Digital Immortality Creators

Digital Immortality Creators are those people creating a digital immortality in the form of one or more of the following:

- Creating digital memories and artefacts pre-death – memory creators;
- Creating a representative avatar pre-death – avatar creators; or
- Creating a digitally immortal persona pre-death that learns and adapts over time – persona creators.

Digital Immortality Recipients

Digital Immortality Recipients are those people encountering digital legacies, such as relatives, friends, lawyers, politicians and religious leaders. These may be of three types:

- Receivers – who receive the memories and artefacts, including representative avatars or digitally immortal persona, authored by the dead person.
- Preservers – who sustain and keep memories and artefacts to create a legacy, representative avatar or digitally immortal persona.
- Mediators – professionals who encounter legacies, representative avatars or digitally immortal persona, such as priests or lawyers.

DIGITAL LEGACY

Digital legacy comprises any information that exists in digital form after death and includes social media profiles, email, online shopping accounts, digital music or photos as well as account information, digital assets and digital property – things that are static once the user has died. (Bassett, 2015, 2017). One of the difficulties with digital legacy is the lack of case law. For example, if an avatar were to commit a criminal or civil offence, but it was created by someone already dead, then it would seem there would be no one to prosecute (Savin-Baden et al., 2017).

Digital commemoration is one area that crosses the boundaries of digital immortality and digital legacy to provide particular commemoration services. These include:

- Memorials and tribute pages hosted on special memorial sites;
- Ceremonies, such as funeral and memorial services in 3D virtual worlds, such as Second Life;
- Solar power headstones with a Quick Response (QR) Code (the matrix barcode) that provides information about the deceased; and
- Digitally mediated funeral practices, such as augmented coffins.

Examples of digital commemoration include websites such as Dead Social that enable users to instruct their Facebook and Twitter accounts to post future updates after they are dead, such as pre-prepared birthday messages. It prompts you to create a social media will, quoting the UK law society – presumably as a means of self-legitimation. While the service is free, Dead Social has distinct enrolment periods and during enrolment, they allow 10,000 users to subscribe to the service.

Digital traces are the traces, or digital footprints, left behind by interaction with digital media. These tend to be of two types: intentional digital traces – emails, texts, blog posts; and unintentional digital traces – records of website searches, logs of movements. Accidental residents of the digital afterlife who leave unintentional traces are seen as internet ghosts or the 'restless dead' (Nansen et al., 2015). The somewhat eerie consequences and impact on recipients are unclear, particularly in relation to ancestor veneration avatars where people are immortalized as avatars in online roleplaying games (Bainbridge, 2013). The traces may be intentional creations pre-death, or unintentional for the dead but intentional by those left behind.

DIGITAL IMMORTALITY AND VIRTUAL HUMANS

The idea of creating digital immortality is one that appears to raise issues and concerns by those in the field of software development. Maciel (2011) argues that seven issues reflect software developers' concerns about death:

respectful legacy, funeral rites, the immaterial beyond death, death as an end, death as an adversity, death as an interdiction and the space required by death. In later work, Pereira and Maciel (2013) explored the beliefs around death within software development and found that religious and moral values affected sensitivities to the personification of death, which, in turn, affected design solutions. Studies such as this illustrate how cultural, political and religious beliefs can affect the technical landscapes around the design of digital immortality.

Companies such as Eternime (2017), Lifenaut (2017) and Daden Limited represent a more recent move towards the creation of an adaptable digital presence. These companies are seeking to build virtual copies of people. If using Eternime, the individual is expected to train their immortal prior to death through daily interactions. According to the company, data are mined from Facebook, Fitbit, Twitter, e-mail, photos, video, and location information, and the individual's personality is developed through algorithms through pattern matching and data mining. Lifenaut works on a similar principle, enabling people to create mind files by uploading pictures, videos and documents to a digital archive. Data are managed and mapped through tagging, time and place. It also enables the user to create a photo-based avatar of themselves that will speak and, according to the company, learn from the conversations that the user has with them.

The work by Daden and University of Worcester and University of Warwick (Savin-Baden and Burden, 2018) in the UK is to create a virtual human, which is a digital representation of some, or all, of the looks, behaviours and interactions of a specific physical human. This prototype system contains relevant memories, knowledge, processes and personality traits of a specific real person, and enables a user to interact with that virtual persona as though they were engaging with the real person. The virtual persona has the individual's subjective and possibly flawed memories, experiences and view of reality. The information provided by the individual is in a curated form and, as such, only data the individual chooses to share and be included in the virtual persona has been used so far. Whilst the project is focused around the use of such persona within the sphere of corporate knowledge management, the mere existence of the virtual persona immediately raises issues around its digital immortality.

There needs to be an understanding of what seems to be the emergence of different ways that 'creating' digital immortality results in digital legacies and traces, and the impact this may have on both recipients and wider society (Harbinja, 2017). What is evident, however, is that it is possible to see distinctions already in those who create digital immortal personas and those who are affected by receiving them after a loved one dies.

CREATING DIGITAL IMMORTALITY

At its most basic, digital immortality merely requires code and data. Digital identities are data, which can be added to and updated (and even forgotten), and an application built from code with a set of rules (which may themselves be data) which enables the interaction between that data and the real world – but, hopefully, something more than just a simple auto-responder like a Twitterbot (Dubbin, 2013). This section explores how this code may then be used to interact with the real world so that the digital immortal can present itself in various ways, although being only a manifestation. Further, and as discussed elsewhere in this book, there is also no reason why there should not be more than one copy of the code (and data) running instant digital clones.

Figure 12.1 provides a simple overview of a potential digital immortality system, drawing on elements discussed in more detail in Chapters 3 through 6.

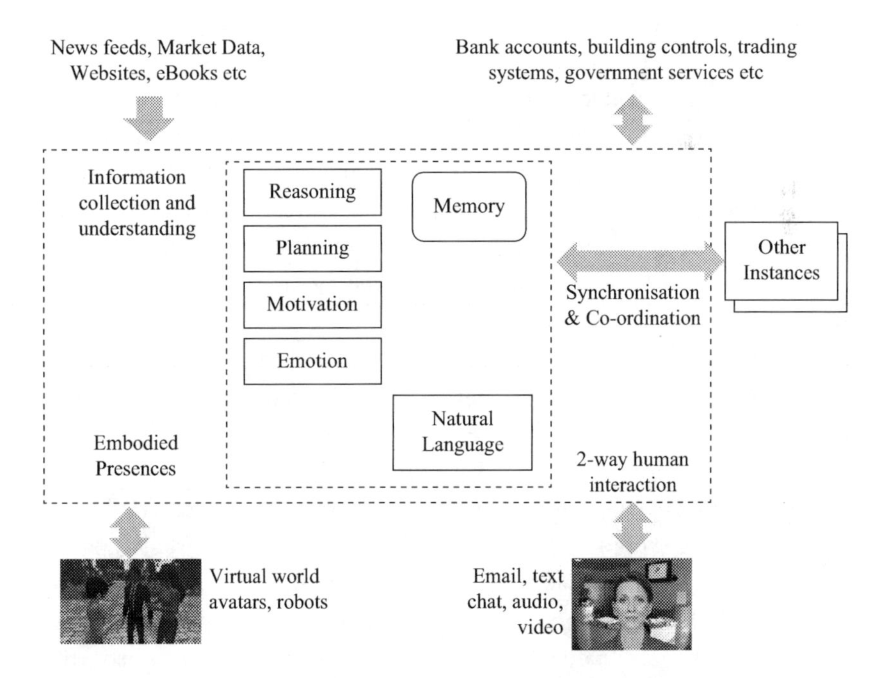

FIGURE 12.1 An overview of a potential virtual humans-based digital immortality system.

A central core manages memory, reasoning, motivation, planning and emotion. The digital immortal can 'read' a variety of information sources and has two-way access to a range of real-world systems. As with any virtual human, it can potentially embody itself in virtual worlds (as an avatar), and possibly in the physical world (via a robot), as well as through a 2D interface, or an email, social-media, aural, chat or Skype presence. It has a natural language understanding and generation facility to enable two-way communication with people (and other virtual humans and digital immortals) in the physical world, and it potentially synchronizes its knowledge and activities between multiple instances of itself.

The most important implication of this code/data existence is that digital immortality becomes, in effect, just a hosting plan (Burden, 2012); as long as you (or your digital immortal) can keep paying the hosting company to run you (and probably keep a backup), you live forever.

Given that the information technology landscape changes over time, there are inevitably also longer-term issues about how the digital identity can be migrated between operating systems, media, computer languages and database standards as they change and evolve – think of moving a program from Cobol or Fortran on a mainframe computer to Objective C on iOS.

What turns this collection of code and data into a digital immortal is the intent (to continue a deceased person's presence and influence) and the illusion that it then creates in the outside world. The important issue here is that, like almost any virtual human, it need *only* create an effective *illusion.* It does not need to create 'consciousness' – although the difference soon moves into the realm of the work of Moody, Dennett, and Chalmers on Zombies (Dennett, 1995).

There are broadly two approaches for creating the data for a digital immortal: manual and automated. In a manual process, the digital immortal might be created by the subject explicitly having a 'dialogue' with a digital immortality 'wizard' over many days, weeks or years. In an automated process, the data would be harvested by tracking real-world interactions of the subject (their emails, voicemails, blog posts, global positioning traces, bank transactions), and the parameters and rules created by modelling their decisions and responses and gradually refining those models (e.g., through neural net or genetic programming approaches) until they accurately reflected the subject's true actions. Any successful system is likely to use a combination of the two. If the subject is already using a virtual assistant or virtual life-coach, then the process of creating the digital immortal could become a background task of the everyday interactions with that virtual agent.

Once a digital immortal is created, it is important to consider how it would interact with the world, and how others might interact with it. Four key areas are identified:

- Passive updating;
- Interacting with systems;
- Interacting with people; and
- Interacting with the physical world.

Passive Updating

The digital immortal can readily collect information about the world, such as 'reading' websites and RSS feeds, reading emails and examining data feeds. In fact, as a digital immortal, it should inherit the email account, bookmarks and RSS feeds of its subject. There are already applications (e.g., Recognant – http://www.recognant.com/) which will extract the key ideas from webpages and RSS feeds, or that can identify trends and outliers in data. Chapter 3 has also described how techniques, such as video analytics and speech recognition, would enable the digital immortal to harvest video and audio information as well. All of this can help the digital immortal's episodic, semantic and even procedural memories evolve, rather than atrophy, as at the time of the subject's death.

Interacting with Systems

Going beyond mere passive updating, by interacting with systems through established Application Programming Interfaces (API) in a two-way fashion, the digital immortal is able to make queries, post messages or request changes in other computer systems. Such interactions may range from simply posting to social media to conducting financial transactions through an online bank or broker. At present, the biggest barrier may be when a 'prove you are not a robot' Captcha is encountered, which is a deliberate attempt to block bot access – but there may be both procedural and creative ways to circumvent this for a digital immortal.

Interacting with People

Chapter 6 has already extensively described the ability for a virtual human to converse with a physical human through natural language, and it can be assumed that a digital immortal would encompass all of this capability, along with the capability to digitally present itself as shown in Chapter 3. The emotional and ethical impact that relatives, colleagues and friends may experience from interacting with the digital immortal has been discussed earlier in this chapter. Obviously, when the digital immortal interacts with people to whom it is not known, then the interaction would be devoid of any sense of strangeness, which is important for the next consideration.

Interacting with the Physical World

In order to interact with the physical world, the digital immortal does not need a physical manifestation. Increasingly, the physical world is being controlled by systems, from home lighting to cars. Thus, if the digital immortal can access and control systems, it can also control the system-connected parts of the physical. At a macro-level, if the digital immortal is controlling the funds and companies that its subject owned, its effect on the physical world through its human 'agents' (employees) could be immense. Giving the digital immortal embodiment within the physical world as a robot or android is almost a sideshow. At a more mundane level, there are many sites around the web, for example, Amazon's Mechanical Turk (https://www.mturk.com/) or People Per Hour (https://www.peopleperhour.com/), where users (human or computer) can post tasks to be done by physical humans – and which would enable the digital immortal (or virtual human) to extend its capabilities in both the digital and physical worlds.

CONSENT AND DIGITAL IMMORTALITY

When undertaking research in this area or creating digitally immortal persona for someone else, careful consideration of informed consent is required. The need for informed consent will include discussions about the future portrayal and representation of data, since participants are consenting to the widespread presence of data in the public domain beyond the lifetime of the researcher and participant. Trustworthiness, the process of checking with participants the validity of data collected and agreeing on data interpretations should ensure research accountability, integrity and rigor. Permission will need to be gained from significant others when data collection involves the data of those already deceased. It may also be important to avoid anyone with complicated grief (a chronic, heightened state of mourning), which could be assessed by using The Inventory of Complicated Grief (Prigerson et al., 1995). Additionally, the researchers should work with a bereavement team to ensure harm will not come to those interviewed who are grieving.

CONCLUSION

The emotional, social, financial and business impact of active digital immortality on relations, friends, colleagues and institutions remains an area that is under-researched. Issues of preservation and privacy issues, and the legal implications of a presence on-going beyond the autonomous control of the mortal presence remains both an ethical and legislative conundrum. A key concern for any form of digital immortality will be to maintain its own integrity. At the most basic level, this will be to ensure that it has the hosting environment (public or private) on which to operate. Finally there is a considerable difference between the creation of a digital immortal, which is the hobby project of a programmer, and the digital immortality, which is the legacy project of a political or religious leader or of a global business leader or billionaire entrepreneur.

REFERENCES

Bainbridge, W. (2013). Perspectives on virtual veneration. *Information Society, 29*(3), 196–202.

Bassett, D. (2015). Who wants to live forever? Living, dying and grieving in our digital society. *Social Sciences, 4*, 1127–1139.

Bassett, D. (2017). Shadows of the dead: Social media and our changing relationship with the departed, Discover Society. Available online http://discoversociety.org/2017/01/03/shadows-of-the-dead-social-media-and-our-changingrelationship- with-the-departed/

Birnhack, M., & Morse, T. (2018) Regulating access to digital remains – research and policy report. Israeli Internet Association. Available online https://www.isoc.org.il/wp-content/uploads/2018/07/digital-remains-ENG-for-ISOC-07-2018.pdf.

Bowlby, J. (1981). *Attachment and Loss* (Vol. 3). New York: Basic Books.

Bridge, M. (2016). Good grief: Chatbots will let you talk to dead relatives. *The Times*. Available online https://www.thetimes.co.uk/article/27aa07c8-8f28-11e6-baac-bee673517c57.

Brubaker, J., & Callison-Burch, V. (2016). Legacy contact: Designing and implementing post-mortem stewardship at Facebook. In *Proceedings of the ACM Conference on Human Factors in Computing Systems* (pp. 2908–2919). Santa Clara, CA: ACM.

Burden, D. J. H. (2012). *Digital Immortality*. Presentation at Birmingham, UK: TEDx.

Cuminskey, K., & Hjorth, L. (2018). *Haunting Hands*. Oxford, UK: Oxford University Press.

Dennett, D. C. (1995). The unimagined preposterousness of zombies. *Journal of Consciousness Studies, 2*(4), 322–326.

Dubbin, R. (2013). The rise of twitter bots. *The New Yorker*. Available online http://www.newyorker.com/tech/elements/the-rise-of-twitter-bots.

Eternime (2017). Available online http://eterni.me/

Hallam, E., & Hockey, J. (2001). *Death, Memory, and Material Culture*. Oxford, UK: Berg Publishers.

Harbinja, E. (2017). Post-mortem privacy 2.0: Theory, law, and technology. *International Review of Law, Computers & Technology, 31*(1), 26–42.

Kasket, E. (2019) *All the Ghosts in the Machine: Illusions of Immortality in the Digital Age*. London, UK: Robinson.

Klass, D., Silverman, S., & Nickman S. (Eds.) (1996). *Continuing Bonds: New Understandings of Grief*. Washington, DC: Taylor & Francis Group.

Kubler-Ross, E. (1969). *On Death and Dying*. New York: Macmillan.

LifeNaut Project. (2017). Available online https://www.lifenaut.com/.

Maciel, C. (2011). Issues of the social web interaction project faced with afterlife digital legacy. In: *Proceedings of the 10th Brazilian Symposium on Human Factors in Computing Systems and the 5th Latin American Conference on Human-Computer Interaction* (pp. 3–12). ACM Press.

Maciel, C., & Pereira, V. (2013). *Digital Legacy and Interaction*. Heidelberg, Germany: Springer.

Nansen, B., Arnold, M., Gibbs, M., & Kohn, T. (2015). The restless dead in the digital cemetery, digital death: Mortality and beyond in the online age. In C. M. Moreman & A. D. Lewis (Eds.), *Digital Death: Mortality and Beyond in the Online Age* (pp. 111–124). Santa Barbara, CA: Praeger.

Savin-Baden, M., & Burden, D. (2018). Digital immortality and virtual humans paper. *Presented at Death Online Research Symposium*, University of Hull, August 15–17.

Savin-Baden, M., Burden, D., & Taylor, H. (2017). The ethics and impact of digital immortality. *Knowledge Cultures, 5*(2), 11–19.

Sofka, C., Cupit, I. N., & Gilbert, K. R. (Eds.) (2012). *Dying, Death, and Grief in an Online Universe: For Counselors and Educators*. New York: Springer Publishing Company.

Steinhart, E. C. (2014). *Your Digital Afterlives*. Basingstoke, UK: Palgrave MacMillan.

Stroebe, M., & Schut, H. (1999). The dual process model of coping with bereavement: rationale and description. *Death Studies, 23*, 197–224.

Tonkin, L. (2012). Haunted by a 'Present Absence'. *Studies in the Maternal, 4*(1), 1–17.

Walter, T. (1996). A new model of grief: Bereavement and biography. *Mortality, 1*(1), 7–25.

Walter, T. (2017). How the dead survive: Ancestors, immortality, memory. In M. H. Jacobsen (Ed.), *Postmortal Society. Towards as Sociology of Immortality.* London, UK: Routledge.

CHAPTER 13

Futures and Possibilities

INTRODUCTION

This book has so far focused on the virtual human technologies which are available currently or will be available by 2020. With existing technologies, it is possible to create some form of proto-virtual human, a virtual humanoid which can exhibit some of the characteristics of a physical human, and may even deceive people in some areas, but which, as a holistic digital copy of a physical human, falls short. However, over the next couple of decades, virtual humans are likely to emerge that are to most intents and purposes indistinguishable from physical humans when encountered within virtual spaces.

This chapter considers what the impact of such virtual humans might be, and what significant other developments might take place, within three successive time-frames:

- 2018–2030,
- 2030–2050, and
- 2050–2100 and beyond.

The chapter will then examine how the three main challenges to the developments of virtual humans identified in Chapter 2 might be addressed in order to move from virtual humanoids to 'true' virtual humans, and even to virtual sapiens.

FUTURE CAUTION

Before considering such future scenarios, though, it is important to be cognizant of the fact that recent history is full of technologies and ideas that seemed to offer promise and which then almost disappeared without trace. Supersonic air travel, regular journeys to the Moon and Mars, videodiscs and hovercraft are just some examples. In some cases, the concepts were right but the technology was just not ready, such as Apple's Newton with an iPhone running Siri, others suffered from changing political and budgetary priorities (space travel), others were made obsolescent by rapidly emerging alternate technologies, for example, DVD taking over from videodiscs, the cloud taking over from both, and others were just not maintainable from an economic, environmental or safety perspective, for example, Concorde. Others continued within a small niche (e.g., hovercraft) but never made the wider impact that was once imagined. Could the same fate befall virtual humans?

The Gartner Hype-Cycle

As with almost every technology, virtual humans are located on what analyst Gartner called the Hype Cycle (Linden and Fenn, 2003). This suggests that technologies evolve through four broad stages:

- A rapidly rising slope after a Technology Trigger to a Peak of Inflated Expectations;
- An equally rapid plunge into a Trough of Disillusionment;
- A long slow climb up a Slope of Enlightenment;
- Until eventually reaching a Plateau of Productivity.

Gartner has been tracking many of the component elements and manifestations of the virtual human (e.g., artificial intelligence, natural language, speech processing, computer vision, machine learning, virtual assistants) on their hype-cycle for many years. What is evident is that it is not even a smooth progression along this curve. Instead, the curve often has a fractal nature, with each new step in technological development going through its own mini-hype cycle and progressing just a small way along the overall curve towards an ultimate goal on the Plateau. Indeed, a technology can often be seen to be occupying different points along the curve at once, each in a different guise.

With virtual humans, each of the types described in Chapter 8 is probably located on a different part of the curve. The more mature (and simpler) virtual humans, such as virtual assistants, customer service agents and virtual tutors, are probably already climbing up the Slope of Enlightenment; their capabilities (and failings) are reasonably well known, as are the improvements required to make them truly productive. Manifestations, such as virtual personas and digital twins, have only recently appeared on the curve and are heading for the peak of inflated expectations, and some are potentially already over that peak; take, for instance, the media frenzy over digital immortality a few years ago and the subsequent launch and disappearance of many of the start-ups described in Chapter 12.

Virtual humans as a concept encompasses such a broad range of philosophies, practices and ideas that there is little doubt that some versions of virtual humans will (and do) exist. It is those more sophisticated versions, the virtual personas and virtual persons, over which there must be more doubt. The key question is whether 'productive' versions of them can be built with current and near-future technology, or whether this will not occur until Artificial General Intelligence (discussed in more detail below) has been created. A whole new technology trigger will cause another iteration of the hype-cycle, which could render them obsolete, either through the arrival of something like the Singularity, or a move towards more non-human versions of AI, perhaps as a result of ethical issues.

McLuhan's Tetrad

Marshall McLuhan's Tetrad (Figure 13.1) provides a useful way to analyse the impact of any new media, or indeed of a wide variety of digital innovations, such as the virtual human. McLuhan identifies that an innovation can have four broad types of effect (McLuhan and McLuhan, 1992):

- It can Enhance and amplify something that already exists;
- It can render Obsolescent something that already exists;
- It can Retrieve something that has been lost; and
- It can go into Reverse and have negative effects when pushed too far or subverted.

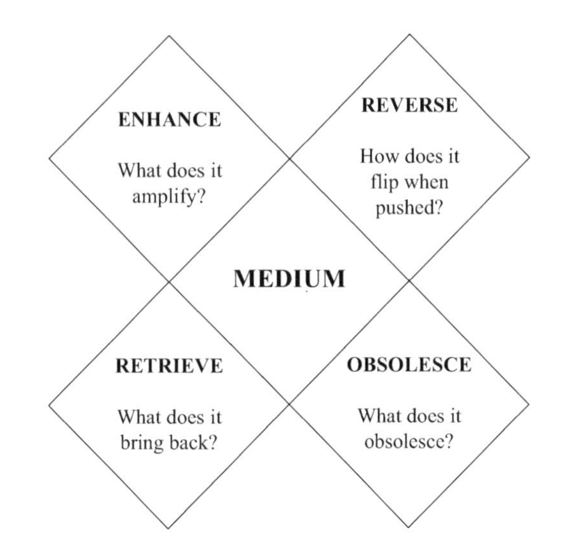

FIGURE 13.1 Marshall McLuhan's Tetrad of media effects.

How might this apply to Virtual Humans?

- Virtual humans amplify and enhance the ubiquity of the humans that already exist and therefore make human-type mental resources unlimited in a similar way that robots make human-type physical resources unlimited. However, the existence of the virtual human in the digital domain means that it is not as limited in resources as physical robots and could be even more pervasive. At a personal level, the virtual persona amplifies and extends the individual, potentially allowing them to be in more than once place at one time. An extreme case of this is in Digital Immortality, which amplifies a person's presence beyond the grave, and, perhaps, renders the concepts of death obsolete.

- General purpose virtual humans – the virtual person – would render many of today's dedicated computer systems obsolete, along with the need to program them. The virtual human could be taught in a similar or accelerated way to do the task.

- The virtual human is already retrieving some of the roles that had been lost or become generally unaffordable in contemporary society, such as the personal assistant (from Siri to J.A.R.V.I.S), or recreating a more human customer services interaction than those provided by an

Interactive Voice Response service on a telephone system, or by adding a virtual tutor to an otherwise impersonal virtual learning system.

- One of the biggest concerns about virtual humans is not, perhaps, the enslaving of the human race to an all-powerful AI, but the possibility of the enslavement of virtual humans by physical humans. Once a virtual person has been be taught how to do something, then perhaps the human owner or controller has no interest in it learning to do something else or exploring its own interests or destinies? This is one of the areas of ethics discussed already in Chapter 8 and explored in science fiction in TV Series such as *Humans*.

Having sounded a few words of caution, what are some of the possible ways in which the future of virtual humans might evolve?

FUTURE SCENARIOS

Taking Stock: Virtual Humans 2018–2030

In the nineteenth and early twentieth century, factories were the home of mass labour. With the rise of automation (machines then industrial robots), factories ceased to be such a mass employer of people. In the mid-twentieth century, the typing pool and related administration departments became the home to large numbers of mid-skill employees. With the arrival of the personal computer, employees undertook their own typing and data entry, and the typing pools disappeared. Instead the call-centre arrived, which, by the end of the twentieth century and into the start of the twenty-first century was, and is, a significant employer of mid-skill labour. Chatbots are beginning to rival human call-centre staff in capability, especially for the simpler tasks, which will result in a reduction in the number of people employed in call centres. Speaking to a virtual agent (or using some other self-service mechanism) will become the norm and may even be mediated by a personal virtual assistant on all media and at home. Text-based bots platforms, such as Facebook Messenger and Slack, appear to offer everyone a virtual assistant who can with help every part of their working day, and with their domestic and personal life as well.

Before moving too far ahead, it is useful to take stock of where virtual humans are now, and how we might expect them to evolve based on currently understood technology. Based on the development presented earlier in this book, Table 13.1 presents the capabilities that a virtual human

TABLE 13.1 Virtual Human Capabilities c. 2030

Digital avatars are visually indistinguishable from physical humans, at least over a Skype call.
Uncanny valley in appearance terms will have been finally crossed.
Have well developed speech recognition and speech generation systems, enabling voice-only conversations.
Can conduct a fluent conversation around a reasonable area of expertise (equivalent to, say, a shopkeeper or office worker 'talking shop') and make a good stab at a very open-ended conversation.
Can regularly pass standard Turing type tests (e.g., the Loebner), assuming that the judges are not setting out to deliberately 'trick' the virtual human.
Can move around a virtual environment as well as an avatar controlled by a physical human.
Can control a robot to give themselves a presence in the physical world.
Have a reasonable facsimile of a human emotion system and can also demonstrate empathy with physical humans.
Can plan activities to achieve a range of goals.
Can access specialist intelligence (automated intelligence, such as driving cars, mining data, playing games) to perform specific tasks which they can integrate with goal and planning activities.
Can learn new information, and, to a lesser extent, new tasks with help from physical humans.
Can access almost any publicly available content on the Internet to answer questions.

could be expected to have within the next 10 years. Table 13.2 identifies the tasks that they may still struggle with.

Even at this level, widely available virtual humans could have a significant impact on a number of different aspects of society. The 'rise of the bots' may not even be restricted to low and mid-skills. Companies such as DoNotPay are already providing virtual humans to enable people to challenge parking fines (Gibbs, 2016), the UK's National Health Service is testing

TABLE 13.2 Challenges for Virtual Humans c. 2030

Being able to:
Respond to long questions with multiple dependent clauses.
Build long-form responses.
Argue.
Understand irony and humor.
Make sense of art in all its forms.
Come up with their own goals.
Think about themselves.
Learn new skills well beyond their current programming.
Explain things or understand explanations of things beyond their current programming.

an AI chatbot developed by Babylon Health (https://www.propelics.com/healthcare-chatbots), and Woebot (https://woebot.io/) is a virtual counsellor employing cognitive behavioural techniques to help address negative thinking.

In the home, virtual humans will also be making their presence felt. Amazon Echo/Alexa, Google Home and other variations on the 'intelligent speaker' or 'smart home assistant' have begun their invasion of our homes. Boy Genius Report (BGR) reported that 10 million Echo systems were expected to be sold by Amazon in 2017 (Epstein, 2017), with TechCrunch reporting Echo lifetime sales being higher at 20 million (Perez, 2017). As Siri and Cortana on mobile phones continue to develop, those living within a developed country will be constantly in reach of a virtual human performing the role of a personal assistant.

Developments 2030–2050: Routes to an Artificial Mind

This book has so far assumed a practical approach to building a virtual human, creating, in particular, the 'mind' element from standard, if complex, computer code. This is what Shanahan (2015) refers to as the 'Engineering AI' approach. There are, however, at least two other approaches to creating a virtual human mind which each have their own vociferous supporters, and which may emerge over a longer timescale. Each will be briefly considered here.

Uploading the Brain

The idea of 'uploading' a human brain to a computer is, to many, a very attractive one. The idea of uploading is part of a broader concept of Substrate Independent Minds (SIM) (Koene, 2012), i.e., that a mind should be able to run independent of the underlying hardware, be that an organic brain or digital computer. Whole Brain Emulation (WBE) is extensively described by Sandberg and Bostrom (2008) and fundamentally consists of translating the detailed brain scan data into a software model based on the behaviour of the neurons and other key parts of the brain. The underlying software provides the generic brain biology/biochemistry/electrical models of neuron and synapse behaviour, and the scan data then details the unique connectivity of the individual brain.

There have been numerous challenges to the whole SIM concept and WBE approach. For instance, Cheshire (2015, 135) states that:

> The suggestion that brain uploading could be achieved safely suggests unbridled hubris. The belief that human identity could be

> faithfully replicated in a machine is possible only within a reductionistic, hence inadequate, understanding of the human person. A hypothetical post-neuron future in silicon could never be more than a collection of inauthentic human representations.

There are also arguments related to the impracticality of disembodied consciousness (e.g., Harle, 2002), but Eth (2013) identifies the Second Life virtual world (discussed in Chapter 7) as being a suitable, if mediocre, environment in which a WBE could exist.

Whilst SIM and WBE are valid directions for research, they do appear to be all or nothing approaches, and with a very long gestation period. In the meantime, a more engineering-based strategy can provide immediate benefits and incremental improvements. Indeed, it may be that engineered virtual humans are what enable uploaded virtual humans to be realized.

The Technological Singularity

Closely linked to the idea of brain uploads is the concept of the Technological Singularity. Borrowing from the cosmological idea of a point where existing conceptions of space and time break down (such as in a black hole), a technological singularity is a point in our history where technological development has been so radical that we can't see beyond it. Shanahan (2015, xv) states that:

> [A] singularity in human history would occur if exponential technological progress brought about such dramatic change that human affairs as we understand them today came to an end ... Our very understanding of what it means to be human, to be an individual, to be alive, to be conscious, to be part of the social order, all this would be thrown into question, not by detached philosophical reflection, but through force of circumstances, real and present.

The Technological Singularity was popularized by Ray Kurzweil in books such as *The Singularity is Near* (Kurzweil, 2005). By around 2050–2060, Kurzweil expects exponential growth to deliver enough processing power to model every human brain on the planet, and with the enabling of that level of 'super-intelligence' the singularity is, to read Kurzweil, inevitable (although it is notable that the first edition of his book in 2005 expected that by 2020, personal computers would match the processing power of the human brain). For Kurzweil, though, humans are still in the driver's seat, with the developments being augmentations to our abilities, rather than being manifest in a single, possibly malevolent, super-intelligence as imagined in several doomsday Singularity scenarios (Yudkowsky, 2008; Boström, 2014).

Kurzweil, and the Technological Singularity itself, have both champions and opponents. Arguments made against his view include those challenging his 'theory of technology evolution' (e.g., Horner, 2008), and those making challenges to his expectation of exponential growth in computing power (e.g., Pacini, 2011). Indeed, Eden et al. (2012) presents a list of 15 challenges to the Singularity concept, and the responses of the scientific community, both supporters and challengers on each. Kurzweil's exponential growth model is linked to a brute force approach to creating super-intelligence, whether by WBE or other means. Whilst this is exemplified in machine intelligence (such as chess computers), more recent advances seem to focus far more on elegance and finesse. Indeed, if the WBE (and nanotechnology) elements of Kurzweil developments are removed, many of the 'impacts' that he identified of the Singularity are readily achievable without such a step-change.

Developments 2050–2100

The competing approaches from AGI (discussed below), Whole Brain Emulation and the Singularity do suggest that 2050–2100 might be the time period in which the development of really complex virtual humans capable of being called virtual sapiens will occur. It would seem reasonable that within the 2050–2100 timeframe, there could be a relatively high abundance of virtual humans, perhaps not complete ubiquity, but somewhere between a high-end smartphone and a car today. This would imply that there could be of the order of 2 – 3 billion virtual humans (energy supplies and global warming allowing), or one for every five biological humans. This does only hold if people find a virtual human useful, otherwise the number will be far lower. So what roles might they play?

Virtual humans will form a part of the economic and social capital. Much has been written in recent years about the 'rise of robots' and the 'end of employment'. If the virtual humans are truly indistinguishable in terms of ability from physical humans, then they would be capable of undertaking a large swathe of current employment tasks, particularly those which are office bound where there is no need for them to take on a robotic manifestation. Frey and Osborne (2017) estimate that about 35% of current jobs in the UK are at high risk of computerization over the next 20 years. By 2050, the number will inevitably be much higher. Virtual humans would almost certainly be cheaper to run than human labour, and probably not demand (or need) so many breaks or holidays. Jobs which require a higher degree of physicality, creativity and, possibly, compassion may be where the penetration of virtual humans is least, but it may only be a matter of degree. Virtual humans will be able to drop into robot bodies as required to perform physical tasks (from building and cleaning to caring making) and will of course also be able to call on industrial and domestic robots to do the more mundane tasks.

If there are virtual human workers, then who will own them? The default assumption would seem to be the employer, but what is to stop potential employees from buying them, probably on some form of credit, and then sending the virtual human out to work for them? They might imbue the virtual human with much of their skill and knowledge, a virtual persona almost, and act as its mentor, perhaps even taking over a task, remotely and momentarily, when it gets beyond the ability of the virtual human. Could that be part of an alternative to the universal basic income model?

Virtual humans will also affect cultural capital. For example, as film production moves into the digital realm, the greater likelihood there is to use virtual humans as actors, initially as extras, then as walk-ons and, finally, as co-stars or even stars. The extras could actually be directed, not programmed, a return to the huge and dynamic crowd scenes of the 1960s. Today's actors are now used to working with colleagues wearing motion capture suits or following the ping-pong ball meant to represent the mouth of a vicious dragon. But perhaps the film of the future is the flip of that, the whole film being made in the digital domain, with digital actors, with just the director, production crew and possibly stars entering the digital world by virtual reality. It need not only be mass entertainment that is affected by virtual humans. There has been a lot of coverage of the rise of sex-dolls (in physical form) and sex-avatars (in virtual forms). The producers of these have got as much interest in the development of virtual human technology as anyone else, creating an emotional and engaging companion, but probably wanting to stop short of giving it too much intelligence or ambition (such as in the film *Ex Machina*). An extension of the virtual gamer is the virtual home companion. Japan's official New Robot Strategy (2015) expects that four out of every five care recipients will have some form of robotic support by 2020, a reflection of the significant investments in the robotic and virtual human technology needed to create the virtual and robotic carers of the future.

THE THREE CHALLENGES

Amidst these possibly utopian scenarios remain the three important challenges identified in Chapter 2 and portrayed in Figure 13.2.

1. How to make a virtual human more human-like in appearance and presentation?
2. How to move from being a 'narrow AI' to an Artificial General Intelligence?
3. Whether it will ever be possible for a virtual human to become truly sentient?

Each of these is now examined, before considering a final scenario in which virtual humans could play their most significant role.

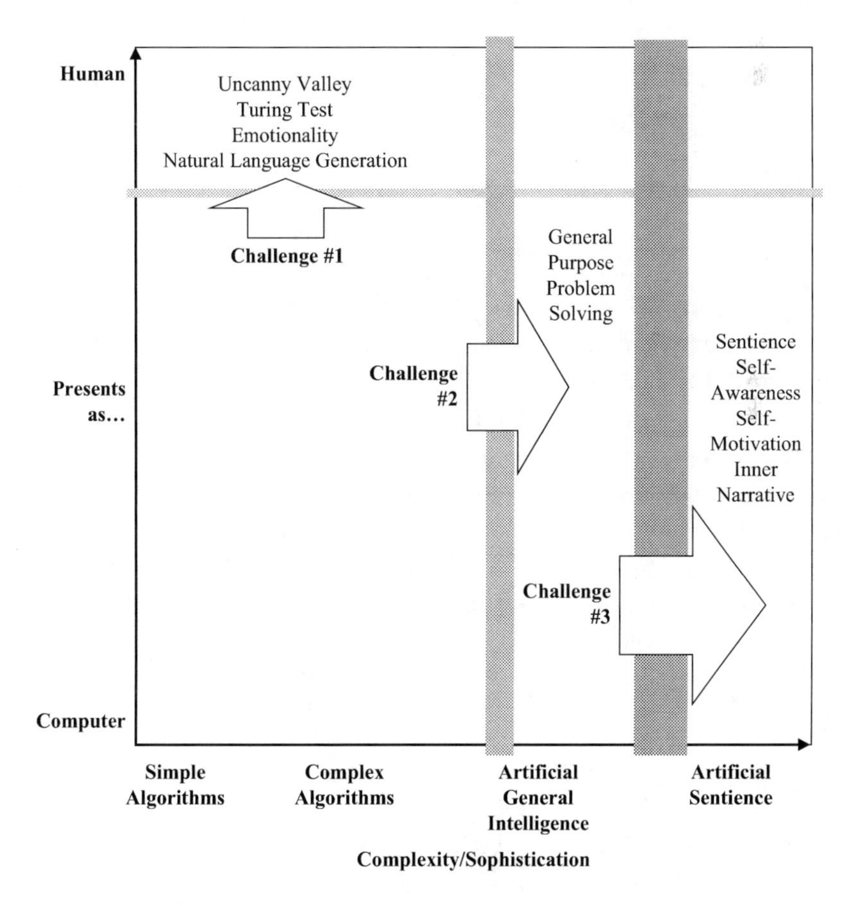

FIGURE 13.2 The 3 challenges of virtual human development.

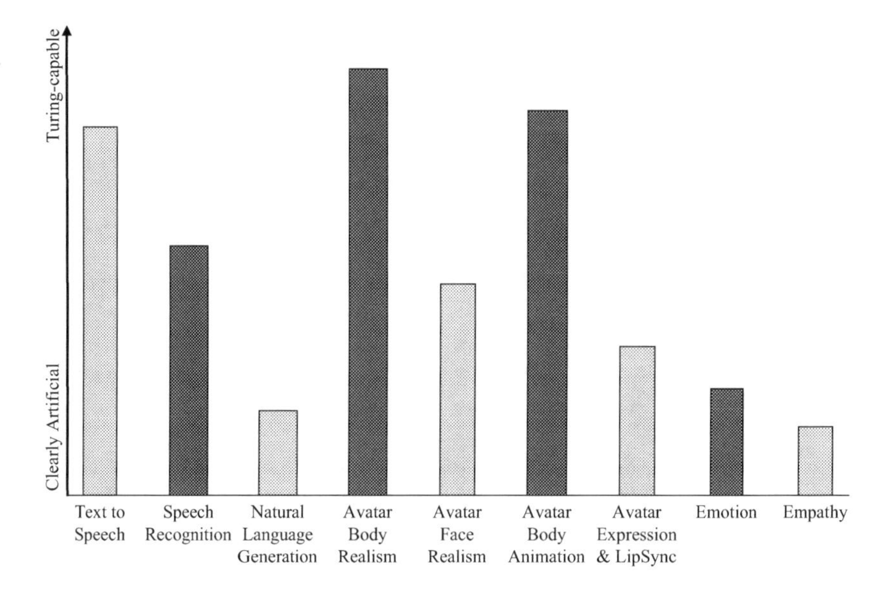

FIGURE 13.3 Maturity of different technologies contributing to 'humanness'.

Challenge 1: Improving Humanness

The creation of 'more human' virtual humans represents Challenge 1 and is likely to be the main focus during the 2020s and into the 2030s. Figure 13.3 provides an overview of the relative maturity of some of the more important technologies involved, based on the discussions in Chapters 3 through 5.

Challenge 1, much of which has been discussed, already is, whilst difficult, probably the easiest of the challenges and is aided by the natural human inclination towards anthropomorphism. People rapidly attribute feelings and intent to even the most inanimate object (e.g., toaster, printer) (Reeves, 1996). So, in some ways, a computer needs to do very little in the human direction for a person to think of it as far more human than it really is. However, this challenge is unlikely to be 100% achieved, particularly in areas such as conversational presentation and empathy, until Challenge 2 has also been addressed.

Challenge 2: Artificial General Intelligence

Artificial General Intelligence (AGI) seeks a return to the 'broad' approach to AI, creating something closer to the science fiction ideals rather than just a system to mine big data or control autonomous vehicles. AGI is seen as more than just the ability to achieve complex goals in complex environments using limited computational resources, and to transfer the resultant learning from one domain to other domains (Muehlhauser, 2013). Goertzel and Pennachin (2007) states that 'what distinguishes AGI work from run-of-the-mill 'artificial intelligence' research is that it is explicitly focused on engineering general intelligence in the short term', in other words it is distinct from both narrow-AI and from theoretical AI research with no practical implementation.

Goertzel (2014) has identified four broad approaches to developing an AGI.

- Symbolic – such as Cyc, SOAR and ACT-R, discussed in Chapter 6.
- Emergentist – a sub-symbolic approach akin to the low-level neuron/synapse model of the brain from which other properties and capabilities emerge. The IBM Blue Brain project (Markham, 2006) is an example of the more computational neuroscience-orientated approach.
- Hybrid – a hybrid of the above two approaches using elements of each in combination; an example is Goertzel's CogPrime, described in Chapter 6, although it lacks the elegance and simplicity of a single approach.
- Universalist – a more theoretical approach based on creating the 'ideal' program given enough computing power to iteratively evolve it. Hutter's AIXI system (2004) is an example (although it bears a frightening resemblance to Douglas Adam's 'nice hot cup of tea' approach to creating the infinite improbability drive).

Reflecting the AGI focus on 'doing things', Muehlhauser (2013) presents four levels of testing for an AGI:

- The Turing Test, as implemented in a 'Gold' Loebner Prize-type competition (see Chapter 5, involving both natural language conversation and audio-visual presentation and understanding);
- The coffee test – just going into a typical (virtual?) house and making a cup of coffee;
- The robot college student test – enrolling and taking classes just like any other student; and
- The employment test – being able to perform an 'economically important' job.

TABLE 13.3 Estimates of Likelihood of Human-Level Machine Intelligence

Researcher	Date of Survey	2020	2028/2030	2040/2050	2075	2150
Baum	2009	10%	50%		90%	
Sandberg	2011		10%	50%		90%
Muller	2012/2013			50%	95%	

Other researchers have added additional levels, including the On-Line Student Test (before the robot college one), the Artificial Scientist Test (after the employment test), and even a Nobel Prize test.

Table 13.3 presents the results of expert surveys into when human-level intelligence might be developed.

- An expert survey by Baum et al. (2011) conducted during the Artificial General Intelligence 2009 (AGI-09) conference based on an AGI capable of passing the third grade (Sotala, 2012).
- An informal survey by Sandberg and Bostrom (2011) conducted at the 2011 Winter Intelligence Conference on developing human-level machine intelligence.
- A further expert survey by Müller and Bostrom (2016) in 2012/13 on a high-level machine intelligence (HLMI) defined as one that can carry out most human professions at least as well as a typical human, so actually beyond the Level 4 test above!

These results suggest that whilst dates slipped from 2009 to 2011, advancements have begun to stabilise, and even improve, since then, suggesting an AGI by mid-century is reasonable expectation, with the end of the century being almost the latest possible date.

In many ways, AGI is a return to the roots of AI research. AGI has brought the focus back to that aspirational future AI – and emphasises that it is by building proto-AGIs, not theorizing about them, that the first real AGIs will be created.

Challenge 3: 2100 and Onwards – Artificial Sentience and Virtual Sapiens

The final step in the journey towards virtual humans is when they move from being 'just' an Artificial General Intelligence-based virtual human to gaining sentience, consciousness, and becoming virtual sapiens. If creating an AGI is probably one order of magnitude, then the greater problem of creating 'humanness', and then creating 'sentience', Challenge 3, is probably at least one order of magnitude greater again, even if it is actually something that one creates or can create or is even possible for an artificial entity to create.

Sentience is something more than intelligence and is certainly beyond what all (or almost all) animals show. It's more than emotion and empathy and intelligence. It's about self-awareness, self-actualization and having a consistent internal narrative, internal dialogue and self-reflection. It's about being able to think about 'me' and who I am, and what I'm doing and why, and then taking actions on that basis – self-determination.

It is a tenet of this book that it is possible to code a bot that *appears* to do much of the things that define sentience. Would that mean one has created sentience (which seems unlikely) – or perhaps sentience has to be an emergent behaviour?

D'Silva and Turner (2012) defines sentient creatures as 'those who have feelings, both physical and emotional, and whose feelings matter to them'. Interestingly, there appears to be more discussion in the literature about sentience applied to animals than to humans. Sentience does not suffer from the problem of having an 'unsentient' state (unlike consciousness), and it is a word that is not liberally applied to all animals. However, there is a move to have dogs classed as sentient (and so probably, by extension, most higher mammals), and popular culture often sees marine mammals and perhaps even cephalopods, such as octopuses, as sentient. So again, it is a scale, perhaps starting from a higher level than consciousness, but still not an absolute.

As discussed in Chapter 8, the lens of personhood is another useful way of considering how sentience might be manifest, including the six attributes for personhood presented by Kristin Andrews (Rutkin, 2016), which could just as well be applied to virtual humans:

- Subjectivity, showing emotion, perspective and point of view;
- Rationality, thinking and reasoning logically;
- Personality, a distinctive individual character;
- Relationships, the capacity to form bonds with, and care for, others, and to be able to accept care in return;
- Autonomy, the ability to make decisions for oneself (with a whole associated debate about the existence and nature of freewill); and

- Narrative self, the sense of having an autobiographically connected past and future, and whether this is qualitatively different from memory.

It is this last element that is perhaps closest to what is needed, along with the associated concept of self-awareness. In creating a virtual human with at least a facsimile of sentience, the ability for self-awareness and meta-cognition will be vital. Researchers and philosophers appear divided as to whether there is ultimately any qualitative difference between human and animal consciousness, sentience, personhood and self-awareness, or whether it is purely a matter of degree, and all creatures are just on different parts of the same spectrum. If the latter then, there should be no issue with virtual humans also occupying a space on that spectrum, and potentially progressing as the technology improves. If the former then, it is yet unclear whether the 'essence' that separates human sentience from animal sentience is something that can be replicated in silicon and bits, or not.

In practical terms though, what difference could be expected between an AGI virtual human and a virtual sapien, given that it may only be a difference of degree? The main features a virtual sapien could be expected to exhibit are listed below.

- Greater self-determination and sense of self. There is no concept of virtual sapiens being owned by anyone without invoking the ghosts of slavery.
- Virtual sapiens will earn their own living.
- Virtual sapiens have the same rights as biological humans.
- Virtual sapiens can build and run their own business and enterprises.
- Virtual sapiens exist as successful artists.
- Virtual sapiens and homo sapiens can form life-long bonds of friendship, and even form civil partnerships, and, according to local laws, marriages.
- Virtual sapiens have the same drive to explore as many humans.
- There will be good virtual sapiens and bad virtual sapiens, perhaps even evil ones.

- Virtual sapiens will be acutely aware of their need for memory, processing power and electrical power, and will take the same steps to protect their access to them as homo sapiens would if deprived of air.

THE ANTHROPOMORPHIC CHALLENGE

This book has taken an unashamedly anthropocentric stance in the consideration of virtual humans - it's in the title after all – but that is not to say that the potential for non-human AIs of equal sophistication is not of interest. However, researchers have a hard time understanding the subjective inner life of our own species, let alone trying to understand that of a dog, bat or octopus. But there is no doubt that trying to create such virtual animals may well provide not only a better understanding of how those creatures operate, and hold up a mirror to our own species, but also suggest new, different or hybrid ways in which virtual 'others' could be created – which may ultimately be better suited to many tasks than a virtual human. As Marenko (2018) points out, creating artificial intelligence could perhaps be focused on developing artificial alien intelligences, although there is no reason why this should not happen alongside the development of virtual human AIs. Indeed, Marenko (2018, 12) describes a concept called FutureCrafting, the reconceptualizing of contingency and the rethinking of uncertainty:

> FutureCrafting is speculation by design, a performative rather than descriptive strategy, whose interventions are designed to prompt, probe, and problematize, to inject ambiguity and even the non-rational and the non-sensical.

In creating even primitive virtual humans, such as the Halo character described in Chapter 8, the authors of this work would embrace such a FutureCrafting approach, exploring what Marenko describes as the 'otherwise' space between 'could' and 'is'. People encountering early future humans are, in many ways, encountering early versions of the future and are forced to consider the possibilities and implications that the creation of such entities might mean. Such encounters may have far more impact than hours of theorizing.

Marenko's concerns with the digital future are two-fold. How do things change when everyone is connected, first in the 'cyberneticization of the world' and a 'technological unconscious', and second, with a consideration of AI in a non-human form, 'moving away from the anthropocentrism that permeates most of the current attitude towards AI'. Indeed, Marenko's 'technological unconscious' sounds reminiscent of Newton's Mindscape in her Mindjammer RPG discussed in Chapter 2 (Newton and Snead, 2016).

There is no doubt that virtual humans would be a key element of a future technological unconscious. Physical humans could be expected to interact and collaborate with a panoply of other physical humans, virtual humans, other AIs and simple systems. Out of this group though, it is the physical humans who could become the outsiders, the incomers, the digital *parvenus*, the out-group. All the others would be able to communicate at a machine level, not needing to convert from an Application Programming Interface to something as slow as speech or text. If the virtual humans and non-human AIs were providing the direction in such a scenario, then would the physical humans be left behind? Would the 'technological unconscious' become primarily that of the machines, not of the physical humans? Would physical humans become the irrelevance?

VIRTUAL SAPIENS: THE FUTURE?

If virtual sapiens do arise sometime in the next 100 or 200 years, then what is their impact likely to be? The arrival of virtual sapiens may not change the way in which lesser virtual humans are used to support education, but it may be that some virtual sapiens choose to take up a career in teaching and work with the other virtual humans to help educate the physical humans. Virtual sapiens would expect to be on a par with, and even superior to, human teachers.

This introduces the debate about the legal status a virtual sapien might have. Many of the arguments are closely related to those already discussed for Digital Immortals in Chapter 12, except that now almost by definition the virtual sapien is acknowledged to be conscious, sentient and have personhood, and as such would expect to have the same self-determination as a physical human. It is around this tension between slavery and freedom for sentient AIs that many of the science fiction stories mentioned in Chapter 2 are based, and an issue that physical humankind may well have to address in the coming century.

A key factor in the impact of virtual sapiens may be how many exist. Does each virtual sapien need a cluster of super-computers to run on, in which case, their numbers may be limited, or are the processing requirements modest to allow significant numbers to be produced?

If their numbers are limited, then, in some ways, their impact may be less than that of more limited virtual humans, who would continue with the more routine roles described earlier in terms of work and play, whilst the small number of elite virtual sapiens could take on a leadership or mentoring role, mediating between physical humans and virtual humans. If the virtual sapiens are verging on the level of super-intelligence, then their relationship with virtual humans is likely to be more challenging, and it is unlikely that they would want to follow a human lead.

If their numbers are large, then human civilization becomes a mixed physical human/virtual human one, possibly with all the issues of prejudice and integration that other multi-cultural states experience. This seems a more likely case since, as discussed, it does not appear to be a given that sophisticated virtual humans will need powerful super-computers. Virtual sapiens could be present in all strata of society, or possibly just in the higher levels given their expected capabilities. Affairs and even marriages between physical and virtual humans are likely to become commonplace, and both sides may even find a way to create virtual offspring. In a 'worst case' scenario though, physical humans will be 'returned' to the more mundane jobs that they lost to automation as a way of keeping them busy and occupied, whilst the virtual sapiens use the available computer processing power for more productive uses.

To conclude, an area that requires particular consideration when considering the future of virtual humans and virtual sapiens is that of space exploration. All of the exploration of space beyond the Moon has, so far, been done by robots, and this is likely to remain the case for the next decade or so. Humans may well take up the reins of inner solar-system exploration after that, but in moving beyond the solar system, it is virtual human-led exploration that seems to be the most likely option. The nearest star system, Alpha Centauri, is 4.2 light years (ly) away. A typical, reasonably feasible, unmanned mission, such as the British Interplanetary Society's Projects Daedalus (Martin, 1978) and Icarus (Long, 2010), have a travel time of around 50 years for a fly-by mission, extending to around 100 years if also decelerating in order to go into orbit. No one has a current, realistic plan for a human mission to any of our closest stars. The craft size would be several orders of magnitude bigger, and journey time probably a lot longer.

Whilst an unmanned mission could be purely robotic, a virtual human crew could provide the flexibility, adaptability and 'humanness' that real success might need. A crew of tens or hundreds of virtual specialists and experts could be assembled. During the long cruise phase, not all of them need to be active, and putting a virtual human in cyber-sleep is a lot easier than human cryogenics! In fact, some could be running at lower clock rates than others, so whilst the engineering crew might experience the voyage minute by minute (and possibly accelerate up to millisecond by millisecond during moments of crises), the artistic contingent might experience the whole 100-year trip as a subjective 14-day cruise.

It is important to note that one of the differences between a well-developed virtual human and a simple robotic program is likely to be the (virtual) embodiment described in Chapter 7. As a result, the spaceship is likely to carry its own virtual world. Some areas may reflect spaces on Earth or other parts of the solar system, others may reflect a best guess at the target destination, others may be data visualisation of known space, and others may be completely fanciful. It is an interesting question as to whether the world would also contain a digital twin of their craft, so that they could plan, and even conduct maintenance by working on the virtual craft and having their actions echoed by physical robots on the physical craft. And would they have a virtual bridge from which to command their spaceship?

If the mission has its sights set beyond its first port of call, then it is likely to use in-system resources to build copies of its own spaceship to send on to other planets. With a human crew, there would be obvious issues in maturing a skilled enough crew to man several new probes, but virtual humans can simply be cloned, or undertake an accelerated nurturing and maturing. A 'von Neuman' process (Freitas, 1980) can then be started, where the first expedition sends out five new expeditions on their own 100-year missions to promising star systems another 5–10 light years distant. Each of those, on arrival, starts building more starships, and sends those out, and so on. A virtual human colonization 'wavefront' would start spreading through the galaxy, although at this rate, it would still take around 500,000 years to reach the galactic centre!

Despite the relative slowness of the exploration, there are three very interesting implications of all this. First, if physical humans do finally discover some form of faster-than-light or jump drive then they will not be exploring *Star Trek* style into a relatively empty and unknown universe. Instead, for almost every system they visit, virtual humans will have got there first! That links into the second issue, what will relations be like between virtual and physical humans? Before any human interstellar travel becomes possible, physical humans may number around 10 billion and operate on half a dozen worlds. Virtual humans could number in the trillions and operate across hundreds of star systems. Virtual humans may feel minimal obligations, empathy or friendship towards their physical brethren, and many could even see physical humans as wasteful, inefficient and something that just gets in the way.

Finally, if the above scenario holds good for homo sapiens, then why shouldn't it hold good for an alien species? Alien civilizations could well have their own 'virtual alien' 'wavefronts' colonising their part of the galaxy. So, if 'first contact' happens, it seems far more likely that it will be between virtual humans and virtual aliens, not, unfortunately, between physical humans and physical aliens. But in this moment in time, before virtual humans start to explore the stars, and given that no virtual alien wavefront has already reached us, then perhaps Fermi's question of 'where are they' (Jones, 1985) remains valid. The key point is that the 'they' are not physical aliens, but virtual ones.

REFERENCES

Baum, S. D., Goertzel, B., & Goertzel, T. G. (2011). How long until human-level AI? Results from an expert assessment. *Technological Forecasting and Social Change, 78*(1), 185–195.

Boström, N. (2014). *Superintelligence: Paths, Dangers, Strategies.* Oxford, UK: Oxford University Press.

Cheshire, Jr., W. P. (2015). The sum of all thoughts: Prospects of uploading the mind to a computer. *Ethics & Medicine, 31*(3), 135.

D'Silva, J., & Turner, J. (Eds.) (2012). *Animals, Ethics and Trade: The Challenge of Animal Sentience.* London, UK: Routledge.

Eden, A. H., Steinhart, E., Pearce, D., & Moor, J. H. (2012). Singularity hypotheses: An overview. In A. H. Eden, J. H. Moor, J. H. Soraker, & E. Steinhart (Eds.), *Singularity Hypotheses* (pp. 1–12). Heidelberg, Germany: Springer.

Epstein, Z. (2017). We finally have an idea of how many Alexa smart speakers Amazon sells. *BGR.* Available online http://bgr.com/2017/06/02/amazon-echo-sales-figures-2017-est/.

Eth, D., Foust, J. C., & Whale, B. (2013). The prospects of whole brain emulation within the next half-century. *Journal of Artificial General Intelligence, 4*(3), 130–152.

Freitas, Jr., R. A. (1980). A self-reproducing interstellar probe. *Journal of the British Interplanetary Society, 33*(7), 251–264.

Frey, C. B., & Osborne, M. A. (2017). The future of employment: How susceptible are jobs to computerisation? *Technological Forecasting and Social Change, 114,* 254–280.

Gibbs, S. (2016). Chatbot lawyer overturns 160,000 parking tickets in London and New York. *The Guardian.* Available online https://www.theguardian.com/technology/2016/jun/28/chatbot-ai-lawyer-donotpay-parking-tickets-london-new-york.

Goertzel, B. (2014). Artificial general intelligence: Concept, state of the art, and future prospects. *Journal of Artificial General Intelligence*, *5*(1), 1–48.

Goertzel, B., & Pennachin, C. (Eds.). (2007). *Artificial General Intelligence* (Vol. 2). New York: Springer.

Harle, R. F. (2002). Cyborgs, uploading and immortality—Some serious concerns. *Sophia*, *41*(2), 73–85.

Horner, D. S. (2008). Googling the future: The singularity of Ray Kurzweil. In T. W. Bynum, M. Calzarossa, I. Lotto, & S. Rogerson (Eds.), *Proceedings of the Tenth ETHICOMP International Conference on the Social and Ethical Impacts of Information and Communication Technology: Living, Working and Learning beyond Technology* (pp. 398–408). Mantova, Italy: Tipografia Commerciale.

Hutter, M. (2004). *Universal Artificial Intelligence: Sequential Decisions based on Algorithmic Probability*. Berlin, Germany: Springer Science & Business Media.

Jones, E. (1985). 'Where is everybody?', An account of Fermi's question', Los Alamos Technical report LA-10311-MS, March 1985. Available online https://fas.org/sgp/othergov/doe/lanl/la-10311-ms.pdf.

Koene, R. A. (2012). Embracing competitive balance: The case for substrate-independent minds and whole brain emulation. In A. H. Eden, J. H. Moor, J. H. Soraker, E. Steinhart (Eds.), *Singularity Hypotheses* (pp. 241–267). Berlin, Germany: Springer.

Kurzweil, R. (2005). *The Singularity Is Near: When Humans Transcend Biology*. London, UK: Penguin.

Linden, A., & Fenn, J. (2003). Understanding Gartner's hype cycles. Strategic Analysis Report Nº R-20-1971. Gartner, Inc. Available online https://www.bus.umich.edu/KresgePublic/Journals/Gartner/research/115200/115274/115274.pdf.

Long, K. F., Obousy, R. K., Tziolas, A. C., Mann, A., Osborne, R., Presby, A., & Fogg, M. (2010). PROJECT ICARUS: Son of Daedalus, flying closer to another star. *arXiv preprint arXiv:1005.3833*.

Marenko, B. (2018). FutureCrafting. Speculation, design and the nonhuman, or how to live with digital uncertainty. In S. Witzgall., M. Kesting, M. Muhle, & J. Nachtigall (Eds.), *Hybrid Ecologies*. Zurich, Switzerland: Diaphanes AG. Available online http://ualresearchonline.arts.ac.uk/12935/1/FutureCrafting_MARENKO.pdf.

Martin, A. R. (Ed.). (1978). *Project Daedalus: The Final Report on the BIS Starship Study*. London, UK: British Interplanetary Society.

Markham, H. (2006). The blue brain project. *Nature Reviews Neuroscience*, *7*(2), 153.

McLuhan, M., & McLuhan, E. (1992). *Laws of Media: The New Science*. Toronto, Canada: University of Toronto Press.

Muehlhauser, L . (2013). *What is AGI?* On Machine Intelligence Research Institute blog. August 11. Available online https://intelligence.org/2013/08/11/what-is-agi/.

Müller, V. C., & Bostrom, N. (2016). Future progress in artificial intelligence: A survey of expert opinion. In V. C. Müller (Ed.), *Fundamental Issues of Artificial Intelligence* (pp. 555–572). Berlin, Germany: Synthese Library; Springer.

New Robot. Japan's Robot Strategy—Vision, Strategy, Action Plan // The Headquarters for Japan's Economic Revitalization. (2015). 90 r. Available online http://www.meti.go.jp/english/press/2015/pdf/0123_01b.pdf.

Newton, S., & Snead, J. (2016). *Mindjammer: Transhuman Adventure in the Second Age of Space* [*Role-playing Game*]. Harlow, UK: Mindjammer Press.

Pacini, H. (2011). Issues in the path to the singularity: A critical view. *Journal of Consciousness Exploration & Research*, *2*(5), 691–705.

Perez, S. (2017). Amazon sold 'millions' of Alexa devices over the holiday shopping weekend. *TechCrunch*. Available online https://techcrunch.com/2017/11/28/amazon-sold-millions-of-alexa-devices-over-the-holiday-shopping-weekend/.

Reeves, B., & Nass, C. I. (1996). *The Media Equation: How People Treat Computers, Television, and New Media Like Real People and Places*. Cambridge, UK: Cambridge University Press.

Rutkin, A. (2016). Almost human? *New Scientist* (p. 17), July 2.

Sandberg, A., & Bostrom, N. (2008). Whole brain emulation: A roadmap technical report. Future of Humanity Institute, Future of Humanity Institute, University of Oxford, Oxford, UK. Available online http://www.fhi.ox.ac.uk/reports/*2008*-3.pdf.

Sandberg, A., & Bostrom, N. (2011). Machine intelligence survey. Technical report, 2011-1. Future of Humanity Institute, University of Oxford, Oxford, UK. Available online www.fhi.ox.ac.uk/reports/2011-1.pdf.

Shanahan, M. (2015). *The Technological Singularity*. Cambridge, MA: MIT Press.

Sotala, K. (2012). Advantages of artificial intelligences, uploads, and digital minds. *International Journal of Machine Consciousness*, *4*(1), 275–291.

Yudkowsky, E. (2008). Artificial intelligence as a positive and negative factor in global risk. In N. Bostrom & M. M. Ćirković (Eds.), *Global Catastrophic Risks* (pp. 308–345). New York: Oxford University Press. Available online https://intelligence.org/files/AIPosNegFactor.pdf.

Glossary

AIML: Artificial Intelligence Mark-up Language, a common language used to create chatbots and/or conversational interfaces

alt-avatar: the creation of an alternative avatar to the one used on a more everyday basis. Often used by staff for social purposes, or to remain unnoticed by students, if they are in-world and wish to be anonymous

anonymity: means that no identifiable data (e.g., name, address, Social Security number) is collected as part of the research study, and, therefore, no identifiable data can be shared

artilect: an artificial intellect, a sentient AI

augmented reality: the live view of a physical world environment whose elements are merged with computer imagery, thus, it places emphasis on the physical, so that information from virtual space is accessed from within physical space. An example of this would be the projection of digital 3D objects into a physical space

augmented virtuality: this is where the virtual space is augmented with aspects from physical space, so there is a sense of overlay between the two spaces

autonomous agent: a software programme which operates autonomously according to a particular set of rules. Sometimes used synonymously with chatbots, but also often used to describe the non-human agents within a simulation or decision-making system

avatar: the bodily manifestation of one's self or a virtual human in the context of a 3D virtual world, or even as a 2D image within a text-chat system

belief-desire-intention (BDI): a common software model for developing intelligent agents and virtual humans where the agent has a set of beliefs about the world, a set of desires to change things and the intention to enact some of those desires

chatbots: software programmes which attempt to mimic human conversation when communicating with another (usually human) user. The Turing Test is a standard test of the maturity of chatbot technology. Chatbots may also be used to control 3D avatars within a virtual world, 2D avatars on a website or exist as participants within text-only environments, such as chat rooms. Chatbots are conversational agents which support a wide range of natural language and extended conversations, rather than just question and answer or command and response

computer vision: the technical approaches which enable computers to see and understand imagery and video

confidentiality: involves the collection of identifiable data about research participants, which is then restricted to particular individuals depending upon the confidentiality agreement

conversational agents: computer programmes which use a natural language rather than a command line (or other) interface to a system, database or programme

digital capabilities: the range of skills and understandings needed to operate digital media and collaborate and share with others, as well as being digitally literate and fluent

digital fluency: the ability to shift easily between and across digital media, often unconsciously, with a sense of understanding of the value and possibilities of their use and function

digital human: a software version of a human which is typically focussed on the size and shape of the human for ergonomic research purposes

digital learning: learning that takes place specifically through digital media, such as virtual worlds, blogs, Tumblr, kik and YouTube, and searching, but in a cohesive way. Digital learning is not about merely acquiring chunks of information from spaces such as Wikipedia; it is learning that is organised

digital literacy: is seen as the ability to assemble knowledge, evaluate information, search, and navigate, as well as locating, organising, understanding and evaluating information using digital technology

digital spaces: those spaces in which communication and interaction are assisted, created or enhanced by digital media

digital tethering: the constant interaction and engagement with digital technology, the sense of being 'always on', 'always engaged'; characterised by wearing a mobile device, texting at dinner, and driving illegally while 'Facebooking'

disciplinarity: all that is seen as central to a given discipline; its pedagogy, values, beliefs, rhetorics and expected norms that are embodied by the academics who guard it

haptics: the use of technology that creates a sense of touch, such as vibration or movement, in order to enhance visual engagement in immersive virtual worlds and other digital 3D environments

identities

bridged identities: identities created to link with other exterior worlds. Such identities might be located through the creation of avatars or using avatars for identity play (playing with avatar identity in ways that are seen as fun and sometimes trite)

frontier identities: these tend to be identities that overlap and overlay on to other spaces but also tend to be twilight identities in the sense that identities sit beside one another and come to the fore when one is required over another

identities on tour: these are dynamic identities, the purpose and point of view of traveller is central

identity tourism: a metaphor developed to portray identity appropriation in cyberspace. The advantage of such appropriation enables the possibility of playing with different identities without encountering the risk associated with, say, racial difference in real life

interstitial identities: metaxis (or metaxy) was used by Plato to describe the condition of 'in-betweenness' that is one of the characteristics of being human

left behind identities: the idea that as we shift and move identities across online contexts; rather than necessarily deleting them, we tend to leave them behind

mapped identities: those identities that tend to be imposed identities, provided by, for example, school reports, external data, job profiles, or human resources records. Thus, these identities are seen as static and objective, even though, in many ways, they are not; they are, in fact, identities mapped on to us by others

networked identities: the idea that identities constructed are multidimensional and complex across overlapping online and offline networks across school, work and spare time, and it is through such networks that individual identities exist and become real

place-based identities: those are identities that are strongly located in relation to the places we inhabit and tend to be relatively stable and ordered

spatial identities: identities enacted through digital media, and each enactment tends to prompt a different kind of performance, invariably guided by the norms, cultures and affordances of both the software and the users of those spaces

infomorph: a software entity with a relatively high degree of intelligence, but possibly not sentient (cf., artilect)

informed: consent involves individuals consenting to participate in research when fully informed of the processes, benefits and harms involved. Inherent in the principle of informed consent is the understanding that the participant can revoke consent, usually up until the study is completed and findings are published

lateral surveillance: the use of surveillance tools by individuals, rather than by agents of institutions public or private, to keep track of one another

learner identity: an identity formulated through the interaction of learner and learning. The notion of learner identity moves beyond, but encapsulates the notion of learning style, and encompasses positions that students take up in learning situations, whether consciously or unconsciously

learning context: the interplay of all the values, beliefs, relationships, frameworks and external structures that operate within a given learning environment

limen/liminal: threshold, often referred to as a betwixt and between state

liminality: characterised by a stripping away of old identities and an oscillation between states, it is a betwixt and between state and there is a sense of being in a period of transition, often on the way to a new or different space

liquid surveillance: the idea that through data flows and regimes of in/visibility everyone is being targeted and sorted

massively multiplayer online roleplaying game (MMORPG): MUVEs in which the focus is on game-based roleplay. The games tend to have some form of progression, social interaction within the game, as well as in-game culture

mixed reality: is seen as a method for integrating virtual and physical spaces much more closely so that physical and digital objects co-exist and interact in real time

multi-user virtual environment (MUVE): a MUVE is seen as a more general reference to virtual worlds than MMORPGs that are games-related, and environments, such as immersive virtual worlds which are not usually seen as games

non-player characters (NPC): characters not controlled by a user within a computer game or MUVE. They may be controlled by a chatbot or by a different sort of software to provide them with planned or autonomous actions and behaviours

open sim: a reverse engineered, open-source version of the code that runs the Second Life virtual world

pedagogical agents: virtual humans used for education purposes

photorealism: the original term refers to the genre of painting based on a photograph, but it is used in digital technology to refer to the highly realistic reproduction of objects, peoples and environments such that they are indistinguishable from images of their real-life equivalents

privacy

- **expressive privacy:** the desire to protect one's self from peer pressure or ridicule in order to express one's own identity
- **informational privacy:** the protection of personal information relating to privacy of finances, personal information and lifestyle
- **institutional privacy:** the protection of personal information and monitoring by organisations, such as governments and banks through CCTV, genetic screening and credit cards
- **social privacy:** ensures privacy on social media sites by using pseudonyms, false accounts and by regularly deleting wall posts, photographs and untagging from other peoples' posts

problem-based learning: an approach to learning where the focus for learning is on problem situations, rather than content. Students work in small teams and are facilitated by a tutor

proxemics: the study of spatial distances between individuals in different cultures and situations

qualia: a term used in philosophy to describe the subjective quality of conscious experience. In virtual worlds, it tends to be used to refer to the illusion and extent of being present in an environment

rez: to make an object appear and instantiate itself within a virtual world

Second Life: a persistent, shared, multi-user 3D virtual world launched in 2003 by Linden Lab. Residents (in the forms of self-designed avatars)

interact with each other and can learn, socialise, participate in activities, and buy and sell items with one another, without any constraints of game play

spatial interaction (model of): it is a model that provides flexible support for managing conversations between groups and can be used to control interactions among other kinds of objects as well

speech recognition: the techniques and technology which enables a computer to recognise and understand human speech

teleport: transferring from one location to another in a virtual world almost instantaneously; this can occur by being offered a teleport by another avatar or choosing to teleport yourself to a new location

text chat: the means of communicating by text message, and, specifically, in immersive virtual worlds by typing a response to another avatar in-world rather than using voice. Text chat may be private, public or in a closed group

text-to-speech: the techniques and technology which can convert text (typically generated by a computer, but possibly also scanned from a page of text) to audible speech

validity: ensures that a study is appropriately designed in order to achieve the original objectives

virtual humanoids: simple virtual humans which present, to a limited degree, as human and which may reflect some of the behaviour, emotion, thinking, autonomy and interaction of a physical human

virtual humans: software programmes which present as human and which have behaviour, emotion, thinking, autonomy and interaction modelled on physical humans

virtual learning environment (VLE): a set of learning and teaching tools involving online technology designed to enhance students' learning experience, for example, Blackboard and Moodle

virtual patients: simulations or representations of individuals who are designed by facilitators as a means of creating a character in a health care setting

virtual reality: a simulated computer environment in an either realistic or imaginary world. Most virtual reality emphasises immersion, so that the user suspends belief and accepts it as a real environment and uses a head mounted display to enhance this

virtual sapiens: sophisticated virtual humans which are designed to achieve similar levels of presentation, behaviour, emotion, thinking, autonomy and interaction to a physical human

voice font: a set of data for a text-to-speech system, which enables such a system to sound like a particular person

voice recognition: the techniques and technology which enables a computer to recognise speech as belonging to a particular individual

World of Warcraft: a highly popular massively multiplayer online role-playing game (MMORPG) developed by Blizzard Entertainment. Players control an avatar to explore locations, defeat creatures and complete quests. The game is set in the world of Azeroth

Index

Note: Page numbers in italic and bold refer to figures and tables respectively.

Printed in the United States
by Baker & Taylor Publisher Services